exotique

French form of

exotic /ɛɡˈzɑdɪk/

adjective

• originating outside a particular place, system, etc.: *those famous publike [sic] Gardens of Padua...are much to be commended, wherein all Exotick [sic] Plants almost are to be seene [sic].*

• attractive, desirable, striking, or glamorous, typically by virtue of being or appearing unusual: *much of Willis's poetry was album verse, with...a silky elegance and an exotic perfume that smack of that very sentimental and artificial school.*

noun

• a plant or animal that has been introduced from another country or climate, and often requires specialist care to thrive: *potatoes were first cultivated as a rare exotic.*

Oxford English Dictionary, 3rd Edition

IN THIS ISSUE

Endpapers: Henri Rousseau, *Tropical Forest with Monkeys*, 1910. John Hay Whitney Collection. Public domain.

Following: *Le Jardin des Plantes. Intérieur de la serre.* 1865. Bibliothèque nationale de France, département Estampes et photographie, FOL-LI-59 [8].

LE JARDIN DES PLANTES.

No. 363.

Intérieur de la Serre.

Bonne d'enfants.

Entrée de la vallée Suisse.

Jardinier.

Gardien.

Villageois.

Le Cygne.

Fleurs exotiques.

Vue du Palais des Singes.

Plantes exotiques.

Imp. lith. Pellerin & Cie à Epinal.

LA+ EXOTIQUE EDITORIAL

LA+ EXOTIQUE delves into many of the themes explored in our most recent issue, LA+ BOTANIC, but does so through design proposals submitted for our 5th international design ideas competition. The competition prompt was to reimagine the forecourt of the Muséum national d'Histoire naturelle in Paris, nestled in the Jardin des Plantes. This site, once a royal garden of medicinal plants from 1635 until the museum's founding in 1793, holds immense historical significance and houses one of the world's most extensive collections of natural specimens. Like those in any 18th- or 19th-century botanic garden or natural history museum, these artifacts were sourced from far-flung places, collected by explorers, armed forces, and colonists as they advanced their missions through global trade networks. However, these institutions are not mere relics of the past; they are sites of active research and experimentation.

The selection of this site for the competition prompts a consideration of how scientific knowledge is produced, and how we come to understand, catalog, and categorize the natural world. The competition challenged participants to envision what a garden in this location, at this time, symbolizes, as well as what it can offer in a time of rapid climate change and extinction, realities that we comprehend through the accumulation of scientific knowledge. As you will see in the following pages, there is a diverse array of responses to this challenge.

In a two-stage process, the jury selected five winners and 10 honorable mentions, all of which are published in full on the following pages. The issue also includes a *Salon des Refusés*, a selection of images from entries that did not place but that were of interest to one or more jury members. The issue concludes with the editor's choice award.

As with all LA+ competitions, we have included an interview with the judges to give a behind-the-scenes look at what types of entries they found most compelling and where there was disagreement. Many noted that numerous proposals critically assessed the colonial legacy of the museum, but they were most inspired by those that offered hope and a way forward for thinking about plant-people-animal relations in a public space rather than those that only offered critique. For example, Paris-based reviewer Catherine Mosbach noted that we cannot simply judge the past with today's knowledge, and no matter how problematic its origins, no place should be treated as a "blank slate." Doing so prevents our ability to encourage knowledge sharing *about* the evolution of our knowledge. Marcel Wilson similarly notes that proposals that only atone for the past neglected to consider the importance of much-needed public space in a dense urban environment, which should be designed to serve multiple needs and publics. The jurors also commented on how the theme *exotique* was approached. Julia Czerniak, Sonja Dümpelmann, and Signe Nielsen all noted the breadth with which entrants interpreted its meaning. Many schemes attempted to assign new connotations to the term by referring to multiplicity instead of using it in a pejorative sense of "othering." Several schemes upturn our understanding of exotic, one by suggesting that humans are exotic on Earth when understood through geologic time.

In our feature essay by LA+ creative director Catherine Seavitt, the meaning of exotic is expanded well beyond the borders of the Jardin des Plantes. The essay begins in the garden with a description of the statue of Comte de Buffon–director of the Jardin du Roi who influenced many naturalists who later worked under the Muséum national d'Histoire naturelle–and then takes us across the Atlantic to Thomas Jefferson, south to Brazil, and back to American shores in the aftermath of WWII. The essay serves as a reminder that the global circulation of plants and animals is only one form of producing the exotic. Just as impactful and undoubtedly relevant to designers are the cultural representations–music, images, films–that color our perceptions about places and people.

LA+ would like to thank all competition entrants who submitted proposals and the jury for their thoughtful deliberations.

Karen M'Closkey
Issue Editor

STRANGE FRUIT;

OR, CARMEN MIRANDA SINGS AT THE WORLD'S FAIR

CATHERINE SEAVITT

Catherine Seavitt is professor and chair of the Department of Landscape Architecture at the Stuart Weitzman School of Design, where she holds the Martin and Margy Meyerson Chair of Urbanism. She is also co-executive director of The Ian L. McHarg Center for Urbanism and Ecology at the University of Pennsylvania and creative director of *LA+ Journal*.

+ LANDSCAPE ARCHITECTURE, DESIGN

"*Exotique.*" An exoticizing of the very word exotic…and, oh so very French! The moniker of this issue's call-for-entries aptly "others" the ordinary and invokes a weird desire. Enlightenment Europe centered itself as pure and superior, and everything else around the globe was either primitive or degenerate. The exoticizing European eye gazed at other plants, other animals, and, of course, other people – and noted their differences. The "naturalists" of the 18th and 19th centuries were sent by various crowns to gather these entities for the royal gaze – they hitched a ride on transoceanic warships, shoulder to shoulder with soldiers, sometimes even sharing a vessel with enslaved people, as they sailed to new lands. Soldiers and naturalists alike were hunters, gathering booty. They returned with the exotic: pressed, flattened, desiccated, skinned – and silent. Bones and fossils reassembled with wire armatures, skeletons without their souls, catalogued in descending order. The war machine produced desire, and back home, that desire was soothed with the artful display of curiosities, from crystals to fossils to plant specimens.

The winning entries of our fifth LA+ international design ideas competition, EXOTIQUE, explore some common themes. Many bring the forecourt of the Muséum national d'Histoire naturelle back home, acknowledging the local terrain and climate, the character of the Parisian ground, anchoring the forecourt through excavation, with a nod to local geology and soils. Indeed, it is a bit more difficult to translocate the whole substrate and not just the fragments of a geology. These cuts in the earth reference the local karst geology, but also paleontology and the fossil record, including both past and ongoing species extinctions. The scientific fact of extinction was established at the museum in the late 18th century – a profound challenge to the human psyche.[1] Loss is felt and acknowledged in many of the schemes, and a changing climate looms large. But several of the competition propositions emphatically celebrate the living, the escaped plants that have found a foothold outside the compartmentalized cabinets and collections inside, even thriving. Our entrants are rethinking the dead and the pressed, instead perceiving the geological ground as vibrant terrain, transforming the crushed, decomposed granite surface of the forecourt into a living, thickened section of soils and seeds. Global plant passengers and stowaways, once considered weedy, are welcomed in these proposals.

A few of the competition renderings, perhaps by chance, include a glimpse of a statue in the Jardin des Plantes, just beyond the museum's forecourt: the Comte de Buffon. Let us take a look at this bronze specter in the garden and consider what he embodies as an anchor of this place.

Georges-Louis Leclerc, later Comte de Buffon, was an industrialist and naturalist – arguably two connected métiers, given their extractive methodologies. He became an appointee of King Louis XV in 1739 as director of the Jardin du Roi (now Jardin des Plantes), a position he held until his death in 1788. Here in the garden, rendered in bronze on a granite base in 1907 by the French sculptor Jean Carlus, sits a weighty Buffon.[2] Clad in voluminous robes spreading across an upholstered chair, the figure leans back confidently with his legs crossed. His right arm is draped over the armrest, fingers nearly grazing the head of a lion, its pelt splayed as a rug under his chair. In his left hand, he delicately holds a live bird, its wings spread, poised to take flight. A naturalist, a hunter, and a lover of birds. Carved into the granite base are his dates (1707–1788) and these bold letters: BUFFON.

Buffon is the author of the monumental 36-volume *Histoire naturelle*, a comprehensive study of minerals, animals, and birds, published from 1749–1788.[3] He is the museum's very intellectual foundation – and his weighty bronze supremacy in the garden reflects his profoundly Eurocentric view of the natural world. Buffon argued that the so-called *new world* of America was an inferior place, with successive generations of species degenerating over time. He declared the animal species of the North American continent as diminutive in comparison with their European counterparts. This, Buffon claimed, was due to the humid air and generally unhealthy climate of America.

Thomas Jefferson, then the former governor of Virginia, bristled and took offense, challenging Buffon in 1785 in his *Notes on the State of Virginia*.[4] He digressed from a detailed naturalist's description of the topography, flora, and mineral elements of Virginia to the more general subject of American animals; specifically, animal size. Query VI of the *Notes* includes Jefferson's rebuttal of Buffon's theory, a lengthy table of animal sizes and weights entitled "A Comparative View of the Quadrupeds of Europe and of America." Jefferson's ultimate defense of Buffon's argument was the evidence of a great *incognitum,* the fossilized teeth and bones of a giant creature that he believed must still be roaming the North American continent. "The skeleton of the mammoth (for so the *incognitum* has been called) bespeaks an animal of five or six times the cubic volume of the elephant, as Mons. de Buffon has admitted. The grinders are five times as large."[5] This creature was, in fact, the mastodon *(Mammut americanum)*, a pachyderm of legend that disappeared over 10,000 years ago. Unknown to both Buffon and Jefferson, the fossils would prove, through the later scholarship of Buffon's student Georges Cuvier, the then-inconceivable condition of a "lost world" and complete disappearance from the Earth: extinction.[6] Species are lost, irreparably. We see this sense of mourning and memorial in the fractured earth of the forecourt proposed by our fourth-place entrants Isabel Yidong Li and Oliver Ziyuan Zhu, "Crumbling Earth," with the names of lost species inhabiting the section of the ground like fossil remains. What is the salve for that impact on the human psyche?

But back to Buffon in the garden: what shall we make of that bird in his hand? It is a living creature, almost in flight, with birdsong. Beyond the flattened taxonomies and stuffed specimens on view in the Muséum national d'Histoire naturelle, there is ephemera and life...the bird represents sound, emotion, connecting the listener to something beyond Paris – a global sound, evoking another place, perhaps a battleground, but also a space of joy. Think of the sounds that might emerge from the proposal of our tied fourth-place winner, Adrien Rousseau. Entitled "The Stone Nests," the project fills the forecourt with a large open scaffold, pierced by irregular concrete totemic columns evoking suspended stalactites, extruded termite mounds, or exaggerated 19th-century Parisian rockwork reaching skyward. Lined with niches and cavities, the columns serve as massive rookeries intended for a variety of bird species. Would the birds indeed come? Buffon, seated heavily in bronze nearby, looks on, with the live bronze bird in his hand.

The 18th-century natural history museum gathered the world, its pressed plants and stuffed creatures arrayed in a building-as-cabinet-of-curiosities, but as colonial expansion progressed, the fascination with these exotic plants and animals encountered in the tropical belt by temperate zone naturalists and their colonizer patrons led to a desire to maintain those species alive in a completely different climate. Wardian cases and new enclosed conservatories of iron and glass were developed in the 19th century to create miniaturized and encapsulated tropical climates for plants, and heated zoos for the menagerie of animals translocated to radically different environs. The ultimate glass house of the Crystal Palace, built in Hyde Park, London by the horticulturalist and "environmentalist" Joseph Paxton to house the Great Exhibition of 1851, made possible the display of live tropical plants and animals within those "walls of glittering crystal that seemed to float in mid-air like a vapour."[7] Prince Albert's opening speech announced the global industrial revolution, and indeed, as its full title suggested, the "Great Exhibition of the Works of Industry of All Nations" would be the first of many such international World's Fair exhibitions. The plants, animals, and minerals on view were no longer objects of exotic curiosity, but objects of industrial desire.

Plants and animals had long been trafficked, but from the 19th to 20th centuries onward, the dialogue of industrial export and import would begin in earnest. Natural histories would now play out as natural resources. Plant-based materials essential to the industrial world of manufacturing were drawn from the tropical belt, including hardwoods, vegetable and mineral oils, cotton, hemp, jute, sisal, and rubber latex. Like Britain, France, Portugal, the Netherlands, and other European nations, the United

States acquired over time a pointillist empire of islands around the Pacific rim. Beginning in 1899, these territorial acquisitions included the Philippines and Guam, ensuring a strategic ability to navigate the world's oceans.[8] The World's Fairs brought these geopolitics to the fore, perhaps most vibrantly at the 1939 New York World's Fair in Queens. That fair was more than simply the presentation of the "World of Tomorrow" – it was the gathering of 60 nations under the specter of a second World War, with a political undercurrent of allied and axis power allegiances on display. Brazil's President Getúlio Vargas was playing the field: would Brazil side with the German and Italian fascist powers, or join the American allies?[9]

The Brazilian pavilion at the New York World's Fair, designed by Lúcio Costa and Oscar Niemeyer, was something of a natural history museum of living things: a tropical garden, an aviary of tropical birds, and a serpentarium of venomous snakes from the toxicology research center in São Paulo, Instituto Butantan. Inside the pavilion were displays of minerals and hardwoods, cotton fibers and paper pulp, all useful resources for the war effort.[10] Culture came along with the resources, with musical programming directed by the composer and conductor Walter Burle Marx.[11] Much of the programming featured the classical Brazilian composer Heitor Villa-Lobos, but it was Carmen Miranda, the Brazilian singer with the exotic and colorful fruit hats, who was the true cultural sensation at the pavilion. During the 1939 World's Fair, she starred in the Broadway musical "The Streets of Paris," and was hugely popular. With Vargas's support, Miranda would become a goodwill ambassador to the United States, symbolizing the accord between the United States and Latin American countries at the meeting of the Pan-American League at the fair. This was an intentional effort to bolster hemispheric solidarity in the face of heightening hostilities abroad.

When the war erupted on September 1, 1939, the Pacific Theater was center stage. American GIs were sent off on ships to exotic island locales, much like the youthful soldiers who shared the vessels with the naturalists of the 18th and 19th centuries. With its longstanding foothold in the Philippines and Guam, the United States continued to cultivate its offshore empire with the gradual acquisition of Alaska, Hawai'i, and the Marshall and Mariana Islands. Much of the staging of the war effort against Japan took place in the Pacific Islands: Micronesia, Melanesia, and Polynesia – a network of military bases creating an American ring of fire. What did these GIs see, hear, and taste of that wartime *exotique*, in that tropical realm?

With the postwar return stateside and to the excitement around the long runup to the statehood of Hawai'i, so much attention turned to that tropical ephemera of sound, indeed birdsong as balm. Enter a new naturalist: the composer, and the 1950s post-

1 Extinction as a natural phenomenon, and an acknowledgment that the Earth is in eternal flux, was first established by the vertebrate anatomist Georges Cuvier in 1796, with the presentation of his paleontological paper "On the Species of Living and Fossil Elephants" at a public lecture on the occasion of the opening of the Institut de France's Academy of Sciences in Paris. Cuvier was then Chair of Comparative Anatomy at the Jardin des Plantes. See Georges Cuvier, "Mémoire sur les espèces d'éléphans vivantes et fossiles," in *Mémoires de l'Institut National des Sciences et Arts: Sciences Mathématiques et Physiques*, vol. 2 (Imprimeur de l'Institut National, 1799), 1–22.

2 Jean Carlus created the plaster for the statue in 1902; the Parisian foundry Joseph Malesset cast the bronze in 1907, and the statue and its plinth were placed in the garden with a formal inauguration on June 13, 1909.

3 Comte de Buffon, *Histoire Naturelle, générale et particulière, avec la description du Cabinet du Roi* (L'imprimerie Royale, 1749–1804). Despite Buffon's ambition to be encyclopedic, the volumes cover only minerals, birds, and quadrupeds.

4 Thomas Jefferson, *Notes on the State of Virginia, Query VI: Productions Mineral, Vegetable, and Animal* (Pritchard and Hall, 1788). This is the first American edition of *Notes*, though printed without Jefferson's permission. The first authorized edition was published in Philadelphia in 1794 by Mathew Carey. The initial draft of *Notes*, a document responding to the 1870 query from the French legation in Philadelphia, François Barbé-Marbois, requesting detailed information about the State of Virginia, was transmitted to Barbé-Marbois in December 1781.

5 Jefferson, *Notes on the State of Virginia, Query VI*, 45.

6 Cuvier used comparative anatomy to argue that the Indian and African elephants were distinct species, and that the mastodon and mammoth fossils ("the great *incognitum*") were distinct from living elephants, and thus represented extinct species. Cuvier, "Mémoire sur les espèces d'éléphans vivantes et fossils."

7 For this apt moniker, see Reyner Banham, "The Environmentalist," *Program: Columbia University* (spring 1962), 58–64. Banham writes: "In that field (of plants) alone, his record is sufficient to guarantee him the title of the first great environmentalist." The description of the Crystal Palace's luminous glass walls is excerpted from George Clinch, "The Great Exhibition of 1851," in *Mayfair and Belgravia* (Truslove & Shirley, 1892), 141.

8 See Daniel Immerwahr, *How to Hide an Empire: A History of the Greater United States* (Farrar, Straus & Giroux, 2019). Immerwahr examines the acquisition of the many territories–including islands, atolls, and archipelagos–that were claimed and governed by the United States from the end of the 19th century until the aftermath of World War II.

9 For a comprehensive reading of the pavilion and its political, cultural, and architectural importance, see Carlos Eduardo Comas, "A Feira Mundial de Nova York de 1939: O Pavilhão Brasileiro," *Revista ArqTexto* 16 (Universidade Federal do Rio Grande do Sul, 2010): 56–97.

BUFFON
1707-1788

war American musical genre of *exotica*. Les Baxter is perhaps the initiator of this little-known style of easy listening music, with his 1952 album *Ritual of the Savage* [*Le Sacre du Sauvage*], the definitive exotica record that influenced so many others.[12] Indeed, exotica is "exotique" – Baxter would dig into the French language and tropical sounds as signifiers, citing Maurice Ravel and Igor Stravinsky as influences on his work. Baxter's lush orchestral arrangements, featuring so-called *tribal rhythms,* layered exotic sounds into his easy-listening lounge music, with songs like "Quiet Village," "Jungle Flower," and "Stone God."

But it was the composer Martin Denny who would truly launch the genre, bringing the post-war naturalists to the sound room. In the 1957 album aptly titled *Exotica*, Denny covered Baxter's "Quiet Village," layering in tropical bird calls and vibraphone rhythms instead of the orchestral strings. Recorded in Waikiki, Hawai'i, Denny captured stylized Polynesiana, and the single was a post-war hit, capturing the rise of Tiki culture and the excitement building around impending Hawaiian statehood in 1959.[13] Denny and others, notably Arthur Lyman and Augie Colon, would develop the genre with novel instruments and natural sounds, a way of transporting the listener to an exotic locale. Denny's successive albums had titles like *Forbidden Island* and *Primitiva*, with soundtracks including animal growls, chants, bird calls, and big cat roars – it was the dangerous jungle life of the naturalist, with the amplification not heard in the museum of curiosities or the bronze Buffon. The critic Wally George of the *Los Angeles Times* would write, "The music has the charm of the islands. ...This is not necessarily a criticism, just an observation: at times you get the feeling you're locked in the Griffith Park bird sanctuary."[14]

Strange fruit. In a post-World War II, early Cold-War era, the same impulse that produced Levittowns and perfect lawns found pleasure in the othering experience of the exotic, of the sonic illusion of tiki lounges, Hawaiian shirts, hula lessons, and the bikini. In many ways, the musical appeal of easy-listening exotica emerged from the Pacific Theater of war, and its appeal to returning servicemen may have been its role as an elixir of sorts for the young men growing older in suburban houses with tiki interiors and shag rugs. These are the midcentury martini-set naturalists, with Mai-Tai cocktails in hand, rather than birds. Denny, himself a World War II vet, admits to the tropical ersatz condition of the exotica genre: "a combination of the South Pacific and the Orient...what a lot of people imagined the islands to be like...it's pure fantasy, though."[15] But is fantasy not the very pseudo-experience engendered by the natural history museum in all its forms, and reflected in the smug posture of Buffon, seated on his lion, listening to his bird?

10 Armando Vidal, *O Brasil na Feira Mundial de Nova York de 1940: Relatório geral, primeira parte* [Imprensa Nacional, 1942]. Vidal, the commissioner of the Brazilian Pavilion, presents a comprehensive accounting of the participants, exhibitors, and programming at the Brazilian pavilion.

11 Walter Burle Marx [1902–1990], the Brazilian composer and conductor, was selected by the Brazilian government as music director for the Brazilian Pavilion of the 1939 New York World's Fair. He is the brother of the well-known modernist landscape architect, Roberto Burle Marx.

12 See Phil Ford, "Taboo: Time and Belief in Exotica," *Representations* 103, no. 1 [summer 2008]: 107–35. See also Francesco Adinolfi, *Mondo Exotica: Sounds, Visions, Obsessions of the Cocktail Generation*, trans. Karen Pinkus [Duke University Press, 2008].

13 For profound scholarship on the American tiki culture movement's commodification of Hawaiian indigeneity, see Stephanie Nohelani Teves, *Defiant Indigeneity: The Politics of Hawaiian Performance* [University of North Carolina Press, 2018].

14 Wally George, "Court of Records," *Los Angeles Times* [September 6, 1959], 24.

15 Martin Denny is quoted by Shuhei Hosokawa, "Martin Denny and the Development of Musical Exotica," in Philip Hayward, ed., *Widening the Horizon: Exoticism in Post-War Popular Music* [John Libbey Publishing, 1999], 75.

Opposite: Bronze statue of Comte de Buffon, Jardin des Plantes, Paris. Public domain, altered.

JURY Q&A

Julia Czerniak is dean and professor of the School of Architecture and Planning at the University of Buffalo. She previously served as associate dean of Syracuse University's School of Architecture from 2014 to 2022. She is educated both as an architect and a landscape architect and her research and practice draw on the intersection of these disciplines. Her most recent design research explores the relationship of design to biodiversity, advancing landscape as a protagonist in remaking and envisioning the complex relationships among animal species. Czerniak's work as a designer is complemented by her work as an educator and writer. Her publications include the books: *Case: Downsview Park Toronto* (1999); *Large Parks* (2007); and *Formerly Urban: Projecting Rust Belt Futures* (2013).

JULIA CZERNIAK

Sonja Dümpelmann co-directs the Rachel Carson Center at Ludwig-Maximilians-Universität Munich where she is professor and chair in environmental humanities. She was previously a professor in the Department of Landscape Architecture, University of Pennsylvania Stuart Weitzman School of Design. Dümpelmann is a historian of urban landscapes and environments in the 19th and 20th centuries. Her most recent monographs are *Landscapes for Sport: Histories of Physical Exercise, Sport, and Health* (ed. Dumbarton Oaks Research Library and Collection, 2022), and the award-winning *Seeing Trees: A History of Street Trees in New York City and Berlin* (Yale University Press, 2019).

SONJA DÜMPELMANN

Catherine Mosbach is a landscape architect and founder of Paris-based design firm mosbach paysagistes and the magazine *Pages Paysages*. Catherine's key projects include the Solutre Archaeological Park in Saône-et-Loire, Walk Sluice of Saint-Denis, the Botanical Garden of Bordeaux, the Shan Shui at the International Horticultural Exposition in Xian, and Lost in Transition Taehwa River Garden Show in Ulsan. She was the recipient of the Equerre d'argent Award with Kazuyo Sejima and Ryue Nishizawa for the Louvre Lens Museum Park and was honored in the Iconic Concept Award category by the German Design Council and Platine Award by INT.design 15th Montreal for Phase Shifts Park in Taichung. The team was honored Firm of the Year 2021 in Landscape and Urban Design by Architecture Master Prize Los Angeles. Catherine was named a Knight of the Legion of Honour by French President Francois Hollande in 2016.

CATHERINE MOSBACH

Signe Nielsen is principal of MNLA (Mathews Nielsen Landscape Architects) in New York. Her body of work has transformed the quality of spaces for those who live, work, and play in the urban realm. Signe believes in using design to promote social equity and community resilience. A Fellow of the ASLA, she has received over 100 national and local design awards for public open space projects and has been published extensively in national and international publications. Signe is a Professor of Urban Design and Landscape Architecture at Pratt Institute and former President of the Public Design Commission of New York City. MNLA's recent award-winning landscape architecture projects in New York City include Little Island, Waterline Square Park, Governors Island, the Rockefeller University Campus, the Whitney Museum, the Edible Academy at the New York Botanical Garden, and the landscape of the iconic TWA Terminal Hotel at JFK.

SIGNE NIELSEN

Marcel Wilson, founder and design director of Bionic, has established the firm as a leader in landscape design. Based in San Francisco, Bionic's diverse portfolio includes infrastructure, coastal adaptation, landscapes on structure, and postindustrial sites. Significant commissions include corporate clients such as LinkedIn, Adobe, and Google, and public realm projects, including large parks, waterfronts, and new urban districts in multiple West Coast cities. The firm has won many international design competitions, including Fort Mason Center Public Realm, the Adobe Creek Bridge in Palo Alto, and the Resilient by Design Bay Area Challenge. Marcel graduated with distinction from Harvard University's Graduate School of Design and was awarded the prestigious Weidenman Prize for Design Excellence.

MARCEL WILSON

+ This competition asked entrants to redesign the forecourt of the Museum of Natural History in Paris in the Jardin des Plantes. The grounds and museums have great historical significance; yet, today, museums grapple with their colonial histories, especially those of natural history and those containing artifacts from faraway lands. Taking the competition results as a whole, what are your general impressions of the entries with respect to our contemporary moment?

JC I'm incredibly impressed by this thoughtful body of work, especially the awarded schemes featured in this collection. Not only does the design imaginary intersect with the impacts of the climate crisis, but the proposals also seize the opportunity to right past wrongs: here, the colonial legacy of botanic collections, the neglect of site specificity, and design based on representations of power and cultural convention over the fluxes and flows of the natural world. I was most taken by projects that proposed future worlds of hope, less so by ones that expressed "desolate" and "grim" landscapes. I don't feel that designed landscapes must remind us of our planetary crisis, as we have reminders all around us. We don't need to add to eco-anxiety.

SD Many entries addressed this moment and the challenges facing today's natural history museums. However, quite a few did not. Perhaps not surprisingly, most entries shortlisted after the first jury-round could be characterized as "Anthropocene projects." Many also complicated this term, which–as scholars have suggested in the last 10 years–reflects a privileged human perspective operating from the Global North and West. Among these Anthropocene projects, I found some that were critical of human activities yet nevertheless struck an optimistic chord, including "The Garden Axis Evolution Project," "The Migration Microforest," and "de|colonize." Others remained rather dark and pessimistic, which in some cases was also given expression in the project title: "Allée of Darkness" and "The Black Axis," for example. Speaking both to the competition prompt and the resulting critical interpretation of the Anthropocene were many projects that addressed geology, geomorphology, soil, and the underground in one or another way, e.g., "The Declassifying Forecourt," "The Stone Nests," and "Liberated Nature."

CM I was surprised by the importance of the colonial scope, particularly the emphasis on knowledge discovered through explorations and expeditions. Participants who adopted this approach placed less value on scientific material. Throughout history, explorers have taken significant risks and even lost their lives to explore unknown territories and expand their spheres of knowledge. This deserves respect. Every era has its dark and light aspects. It is essential to consider each period as a specific moment in human history, with specific knowledge, just as we do for a landscape design strategy in a particular place, without rewriting the specific place, and without rewriting the past with today's level of knowledge. Nowhere is a blank slate. Paying attention to a place's physical and historical context as a niche of biological and anthropogenic hybridization is the basis for engaging in a holistic debate about its future. There are explorers and authors in the broadest sense, such as writers, who are very respectful of the environments they explore. Surveying the world to gain knowledge does not always imply intrusion or disrespectful behavior. This is why a proposal to crystallize colonial behavior in a plant garden competition is not necessarily encouraging in terms of knowledge sharing.

Previous: *Three poppies, two butterflies, a fly and a dragonfly* (1688-98) by Johan Teyler. Original from the Rijks Museum, digitally enhanced by rawpixel, used under CC BY 4.0.

MW Looking across all the entries, the struggle to resolve the complexities of the site, its history, the competition brief, and contemporary ideas of justice and equity was apparent. I also noticed that most entries did not position their proposal in any contextual analysis or in the needs of contemporary Paris. We all know cities are complex and rife with needs for urban life. It wasn't an explicit requirement of the brief, but it should be implied as a basic tenet of design in an urban environment. A radical proposal that makes sense of an obvious contemporary need or contextual observation could be distinct and powerful, especially in an ideas competition.

+ Can you identify specific entries or ways in which you felt the theme *EXOTIQUE* was approached in an interesting or surprising way?

SD *Exotique* is exotic in the English language and, therefore, draws attention to the word, its power, and its meanings. It subverts its original use and meaning, which occurred in the colonial context. There, the word was used in both languages to describe something that was different and coming from a different place or country. Difference, or otherness, was often associated derogatively with what colonizers considered to be uncivilized, uncouth, and strange. We still find "exotic" to be used rather uncritically in texts related to the natural world and in the context of gender relationships, where it also portrays the underlying power structures. Nevertheless, today, the word is also used in a positive sense; and we can see some competition entries, such as "The Garden Axis Evolution Project," embracing this benign exoticism, which inverts its discriminatory meaning by associating it with pluralism. I found it interesting that several competition entries, such as "The Black Axis," "Enigmatic Dimension," "The Foreign Inside Us," and "Crumbling Earth," implicitly or explicitly alluded to humans becoming exotic on Earth. This testifies to the importance of Anthropocene scholarship and criticism in this competition and to the significance of recent Western understandings of more-than-human worlds.

MW Several entries were quite dark in their assessment of the institution's history, the climate, and colonization. They used the competition to propose a subversive design response, a self-inflicted punishment, or an apology for the past. There is currently a window in cultural discourse for reparations for past transgressions, and rightly so. But I would argue that just as monuments to conquerors and abusers are not appropriate for public spaces, neither are counter-monuments, especially in urban areas where the higher and better purposes of public spaces that benefit everyone are so evident and so needed.

JC While discussions about the colonial legacies of botanic gardens have been going on for a while, communicating alternative futures through their transformation–as these winning schemes do through design and visualization–is very exciting. In Paris, we are talking about a garden with over 8,000 species! Ecologists and botanists tell us that the effects of the redistribution of flora by European empires are still visible in global biodiversity today, and catalyzing local biodiversity requires attention to specific biomes. The winning schemes do this in diverse ways, at once critiquing the more colloquial exotics (collections from other places) and proposing new ones. The first-place scheme, "Liberated Nature," is based on freeing the "rigid boundaries" of the site's organization and species, which housed relocated flora from around the world. It also "frees" vegetation from its defined forms and locations and proposes to "allow native vegetation to flourish." This scheme is also smart to anticipate a hotter and drier Paris and posits a sunken garden as a cool sanctuary, which is also a nod to the city's limestone base and quarrying history.

The interpretation of exotic in other schemes is also compelling, playing on its meaning of "foreign collection" but also its more common definition as "excitingly different." Exoticism appears variously as nature-centric, with emergent materials ("Seeds' Ark"); as the detritus of consumer culture ("Garden of Anthropogenic Wonders"), and as mineral columns for nesting birds ("Stone Nests"). I'm particularly taken by the last one.

CM Some of the contributions were unexpected. They introduced *mise en scènes*, placing plants in a context to elicit spontaneous interaction and engagement with all types of audiences. (In France, professionals working inside museums to exhibit scientific or artistic content are called *scenographes*.) The ability to affect or move an audience, in the broadest sense, is one of the main tools and objectives of culture. The competition participants achieved this in a variety of ways. For example, I appreciated projects where the main subjects–plants–were articulated in the environment and not placed as isolated figures or objects. If our main objective is to enhance the accessibility of knowledge to everyone, then it is much easier now than it was in the past. We have many more means available that can help us shed light on the breadth of a vision that should be as broad as possible to encompass the diversity of the human sciences.

+ What did you find most compelling and most disappointing about the proposals?

SN Most winning and honorable mention finalists interpreted *exotique* in very personal terms, broadening the viewer's appreciation of the word's meaning in contemporary society. Layered into these interpretations were schemes that address the challenges of our time, including the impacts of the climate crisis, species extinction, colonialism, and exploitation. This rich interplay of visual history, personal interpretation, and genuine motivation to tackle intractable issues of our time was most compelling.

CM The most compelling schemes were those that took unexpected paths, such as "The Stone Nests," which invested in the smallest scale by developing what appears to be a reference to mitochondria as the tissue of micro-organisms. Or, in a different vein, the very radical "Black Axis," which introduces visitors to an inner world. The most disappointing schemes were those that tried to frame the past with the eyes of today, highlighting inappropriate behavior that is not directly related to the scientific purpose of the museum and garden. Those schemes end up focusing only on the social aspects of our history rather than the scientific purpose of the museum. However, it is possible to present the social context in a way that does not detract from the main objective, which is to educate the public about the environment and the evolution of our knowledge of it while respecting the local environment, both human and biological.

SD I would have expected all proposals to address the colonial context in one way or another. However, this was not the case. I also noticed that some proposals worked little with the actual site conditions and the existing spatial situation. Others did, and these made for some intriguing proposals: notably, "The Garden Axis Evolution Project," "The Parterre of Topiary Anamorphosis," and "The Black Axis." Sometimes, powerful ideas lacked a successful design translation; at other times, designs did not speak that much to the competition prompt. What this design competition also revealed quite clearly is that less is often more. The most compelling entries featured a powerful idea transformed into a spatial language working on multiple levels: emotional and rational.

MW I found the entries that used plants as a medium for creating purpose, space, or horticultural effect to be the most compelling. These entries tested structures, technologies, topography, microclimates, and species. It was disappointing to see how many entries got mired in analysis or a weak idea and never pivoted to a physical design. A competition is a format where every word and image counts. A juror shouldn't have to look for the idea behind a project: it should be immediate, and the words and analysis should deepen the idea and its reading.

JC The competition brief was demanding: a high-quality, sophisticated idea expressed in thoughtful, clear, and powerful writing; a high-quality design strategy; and provocative, hopefully unusual visualizations. Although no project achieved all of these high marks, many came close. In terms of visualization, many perspective drawings were beautiful: portraying misty, naturalized, almost Olmstedian landscapes that seemed appropriate given their focus on emergent, sometimes wild native vegetation. Although these were seductive, I appreciated drawings that helped to convey a scheme's logic, such as the global diagram of collection sites of almost-extinct species in "Seeds' Ark."

In terms of communicating ideas, almost all the schemes critiqued the garden as a whole. This is convenient and perhaps necessary in a competition, but it is important to remember that the garden comprises more than just plants from faraway places. In a more nuanced stroll (or scroll, in my case), one encounters the ecological garden, which is "a preserved enclave presenting the diversity of the natural environments of the Paris Basin" (Jardin des Plantes website). I wish more of the schemes embraced nuance.

+ Given that the site is the Jardin des Plantes, were there any proposals that compellingly worked with vegetation?

SN Following the themes of human destruction of the environment, I was struck by the notion of creating a seed bank of lost species from global biomes, as this appropriately addresses the museum as a research institution. I was similarly moved by the schemes that view the future with hope and positivity by conceiving of nature's ultimate adaptability to overcome drastic alterations to global ecosystems.

JC Most of the winning schemes and honorable mentions dismantled the organizational geometry of the forecourt. Given that so many ideas were based on emergent ecologies, the vegetation appeared rather conventionally. However, two schemes come to mind that challenged the appearance of vegetation: "Gardens of Anthropogenic Wonders" (third place), where the vegetative appears as a series of walls composed of algae tiles generated from collected waste; and the "Parterre of Topiary Anamorphosis." Thank goodness for the editor's choice! I rated this scheme highly, as I found it fresh and intellectually stimulating. The scheme proposes green conical mirrors as optical devices (and topiary-like features) through which optical transformation is enabled. "Vegetation" is used to resolve distortion and make the garden recognizable. The scheme makes me recall some of the design techniques and features of French baroque gardens. Here, the author's use of a conical mirror, where distortion can be corrected by viewing from many angles, seems to challenge the singularity and power of perspectival vantage points, which are quite common in the French landscape. I would have liked to see more of the "Seeds' Ark" vault, which promises visitors the experience of seeing species progress from lab to seed to plant as they burst onto the ground above.

CM The challenge is to coordinate a vision between a building and a historical garden. Proposing a unique interpretation of these two polarities is a challenging exercise. In any case, the design must be in dialogue with the previous stages of the site. Although it might be difficult to construct, "The Stateless Assembly," which was not selected, could be a very phantasmagorical proposal, combining all the levels of the courtyard and museum in an amplified way.

SD Yes, many did, even if they often followed some of the more recent planting paradigms that work with spontaneous plant growth and new ecologies to adapt to climate change. For example, in "de|colonize," visitors distribute plant seeds on the forecourt via the soles of their shoes. "The Liberated Nature Nestled Within a Sinkhole" lets spontaneous vegetation establish itself. And, in "Exploration Grounds," the botanical garden becomes inaccessible to humans: a site for plants to develop on their own. Conversely, in "The Migration Microforest," a test plot is established on the museum forecourt for the identification of climate-resilient arboreal species; and the "Digital Garden" brings extinct plant species back on giant LED screen walls. "The Declassifying Forecourt" uses herbarium plants as a starting point to determine the quantity and type of soil distributed on the museum forecourt.

Sonja Dümpelmann

+ Can you identify specific entries where you felt that history was interestingly approached?

Most entries that took the competition prompt seriously and reflected upon the site were related to the site's historic colonial entanglements. As mentioned above, it struck me how many entries were preoccupied with deep time, i.e., geological and earth time, rather than only with human time and history. Several entries—for example, "The Garden Axis Evolution Project," "The Parterre of Topiary Anamorphosis," "The Foreign Inside Us," and "The Black Axis"—also worked with the region's garden history. For me, projects that explicitly revealed the entanglements of colonial with site/landscape design histories were among the most successful. I therefore very much appreciated entries such as "The Garden Axis Evolution Project" (which did not place in the competition) and "The Parterre of Topiary Anamorphosis" (editor's choice award). "The Stone Nests" (fourth place) is much more tongue-in-cheek; and, while it was perhaps not too concerned with the space for which it was designed, I enjoyed its reference to 19th-century Parisian use of concrete in public parks and gardens.

Catherine Mosbach

+ Your firm has entered and won numerous competitions that have become your signature projects, such as the Bordeaux Botanic Garden and Taichung Central Park (discussed with you in our previous issue, LA+ BOTANIC). Would you have liked to enter this competition?

I am not sure I would have taken part. A capital like Paris has strong constraints on public spaces, which leaves little freedom for unusual and unexpected worlds. When I enter competitions with my team, I need to be able to invest myself freely. Otherwise, nothing will be learned. A landscape project should be designed to move the relationship between us and the wider environment. However, it is important to offer ambitious programs to young professionals starting out. Sometimes, incongruous questions can lead to very pertinent proposals. If the Bordeaux Botanical Garden curator hadn't asked me to represent the natural environments of the greater

Aquitaine basin in 2ha, I wouldn't have had the imagination to propose this concept. We were one of three landscape architecture studios in the Bordeaux competition. The other two teams presented "knowledge objects" to the public, as if in front of a "scene" or painting. Our proposal placed the public in the middle–i.e., on the stage and in the environment. Sometimes, asking the impossible can spark creativity. It is essential to nurture the younger generation's creativity.

Signe Nielsen + Marcel Wilson

+ You both have design practices based in the US, where the competition culture for built work tends to be conservative and commissions are often granted to more established firms. Do you see value in entering competitions in general and ideas competitions specifically?

MW There is undoubtedly value in entering well-run competitions, but competitions for landscape architecture commissions are declining in North America. It is difficult to pinpoint one reason. Philanthropy often plays a role in sponsorship, and that source of support has migrated away because of the perceived loss of control over outcomes inherent to design competitions. Some owners and governments are uncomfortable with this. I also think there is a general lack of expertise in framing, staging, and administering a competition process that is attractive to top talent. Starting in about 2005, competition advisors began overleveraging firms and asking for more and more. At some point, most firms determine that they have better places to apply their resources, leaving only a handful of larger firms to compete repeatedly. Perhaps with a reset in the current economy, competitions will re-emerge with a new purpose, a more balanced set of expectations, and equitable opportunities for diverse practices.

SN Ideas competitions are an opportunity to explore visual narratives (based on a compelling brief) that would be unlikely in a real-world, client-driven scenario. Even though the brief stated that these schemes were to have some basis in the reality of the site, the awarded schemes are probably the least grounded in practicalities. Not all competitions are ideas-based; however, those that are ideas-based are like art, with the creator aiming to evoke new meanings and interpretations of forces and factors affecting our time. Ideas competitions can challenge the viewer's perceptions and understanding of the world as we see it.

Julia Czerniak

+ You wrote the feature essay for our issue LA+ ICONOCLAST, which showcased our second competition. Both ICONOCLAST and EXOTIQUE sites have historical significance, but the history of the site is treated differently in the two briefs: ICONOCLAST's Central Park, NY, was assumed destroyed, while EXOTIQUE asked entrants to respect the grounds and museum façades. Do you have any thoughts on whether that impacted the overall "tone" of the submittals?

I appreciate that the work had to consider the significant historical context of the site. Although the proposals for LA+ EXOTIQUE were more nuanced and refined than those submitted for LA+ ICONOCLAST, they were no less imaginative. In fact, they had more power, as they seemed possible and, therefore, influential and impactful.

As I said then: international competitions serve an important role in the landscape discipline. They showcase the work of talented, often emerging designers and foreshadow the concerns that will engage future professionals. If these bright ideas are the future, then I have a lot of hope for our field.

The Jardin des Plantes and Menagerie in Paris

Paul LeGrand, 1842. Metropolitan Museum of Art, Gift of Lincoln Kirsten, 1979.

Quai
St
Bernard
Buffon
50
44
43
20
20
20

WINNERS

PEIXUAN WU
LIWEI SHEN
JINGYAN WANG

LIBERATED NATURE NESTLED WITHIN A SINKHOLE

As a traditional botanic garden, the Muséum national d'Histoire naturelle forecourt embodies a disciplined view of nature. Rooted in a Western colonial mindset, it comprises relocated flora from across the world. While celebrated for its historical significance, the botanic garden treats nature as a curated display of royal authority and human dominion, ignoring local geology and environmental shifts. Liberated Nature aims to upend this tradition. In stark contrast to the confined, rigid boundaries of the past, the new design allows native vegetation to flourish freely, creating a vibrant, self-sustaining ecosystem.

The proposed design intends to educate the public about local climate and geological history to support a localized, contextualized view of the museum grounds. Lowering the ground level and taking advantage of site features, including areas of known former quarries, creates a moister and cooler microclimate that approximates the region's historical climate. This serves both as a nod to the native ecosystems that were once in this area and as a pioneering effort to address the potential geological calamity of sinkholes in the limestone geology of Paris–a city that, because of the extensive former quarrying operations, is sometimes said to be "built on air." As Paris gets drier and hotter, the cooler microclimate of this sunken garden offers a sanctuary for the public and provides a flexible approach to extreme weather by accommodating changes in water levels from rainfall. The museum's forecourt will evolve from a colonial-era relic to a dynamic educational space. Where human influence once reigned, the power of nature and evolution now takes center stage.

1

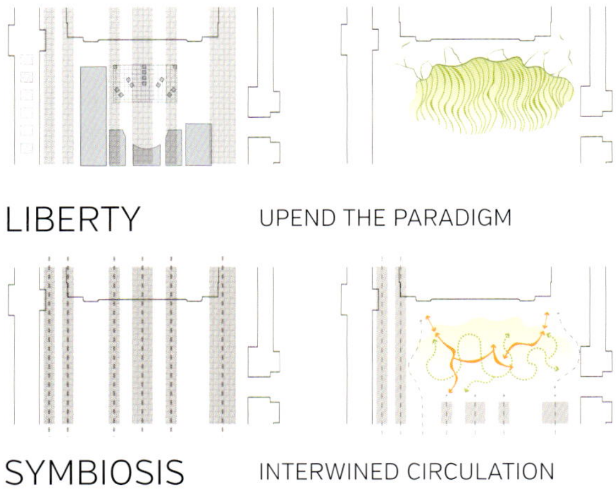
LIBERTY
UPEND THE PARADIGM
SYMBIOSIS
INTERWINED CIRCULATION

±0.00
-2.00
-4.00

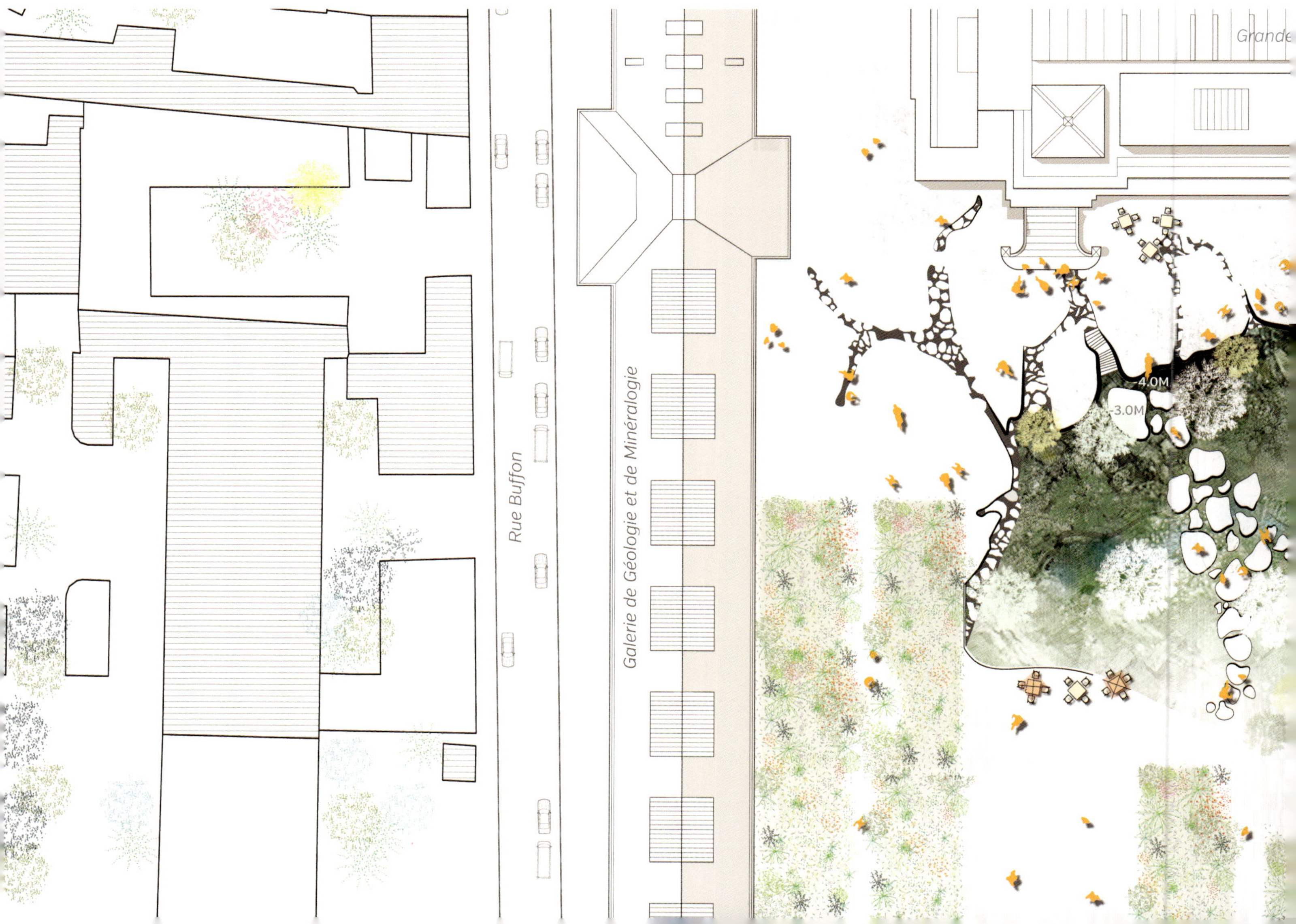
Grande
Rue Buffon
Galerie de Géologie et de Minéralogie
-4.0M
-3.0M

NATURE-BASED DESIGN

YINING ZHANG
LING ZHANG
YUEHUI GONG

SEEDS' ARK

The Jardin des Plantes is home to numerous exotic species brought back from various voyages to places like Senegal, Madagascar, India, North America, and Brazil. Only species with medicinal or economic value were collected and reproduced in the garden, helping establish French colonial power and the order of nature.

With decolonization and the rise of globalism, what was once considered exotic hardly exists anymore. Seeds' Ark reimagines exoticism as diverse and nature-centric, challenging colonial ideologies by creating an underground seed vault library that preserves plants from all biomes, emphasizing places with high concentrations of endangered species. Unlike a seed vault that functions as a secure facility for "backing up" agriculturally significant plants, the redesign of the museum's forecourt is organized to allow living plants—"escapees" from those planted in the vault below—to germinate. In plan, the geometry follows the historic grid form; in section, glass openings in the ground allow ample sunlight to the underground vault, which includes gallery spaces that invite visitors to experience the progression of a living species from lab, to seed wall, to plant, to above-ground exterior spaces where the plants grow vertically out of the bounds of the vault. The seeds from the plants that poke above the forecourt can germinate through wind, landing upon angular landforms and sowing a ground condition that is a more fertile and biodiverse representation of the planet's future.

2

SEEDS' ARK

Senegalia catechu
Zostera marina
Aloe vera
Pinus albicaulis
Rhododendron ovatum
Acer palmatum
Tulipa gesneriana
Vachellia farnesiana
Cirsium fontinale
Helianthus annuus
Thesium ebracteatum.
Coffea arabica.
Tabebuia aurea
Alsophila tricolor
Rosa chinensis
Vanilla planifolia

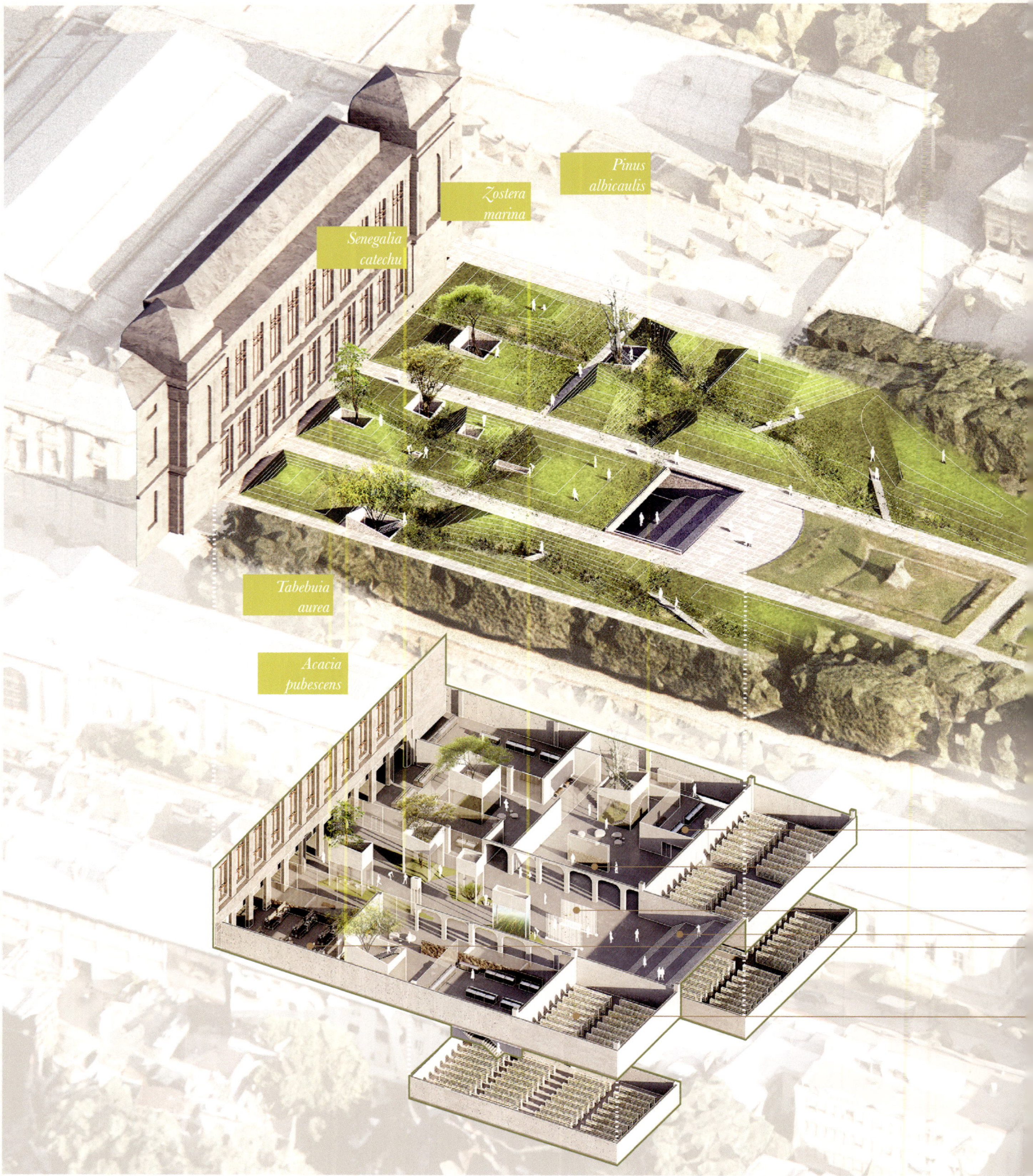

Pinus albicaulis
Zostera marina
Senegalia catechu
Tabebuia aurea
Acacia pubescens

2100
2030
Century
Sowing for Future
Seed Vault for Preservation
French Formal Garden
Planting in Grid
round Seed Garden
orphology Lab
ed Wall
ed Exhibition Curtain
air to Ground Seed Garden
ology Lab
ed Vault
nderground Seed Library

Ground Seed Garden

Seed Wall

Biology Lab

Exhibition Alley

+ Scientific Name - *Rosa chinensis*
Colonized Region Found - East Asia
Reason for Collection - Ornamental Use

+ Scientific Name - *Tulipa gesneriana*
Colonized Region Found - Central Asia
Reason for Collection - Ornamental Use

+ Scientific Name - *Aloe vera*
Colonized Region Found - Africa
Reason for Collection - Medical Use

+ Scientific Name - *Vachellia farnesiana*
Colonized Region Found - Caribbean
Reason for Collection - Ornamental Use

+ Scientific Name - *Rhododendron ovatum*
Colonized Region Found - East Asia
Reason for Collection - Ornamental Use

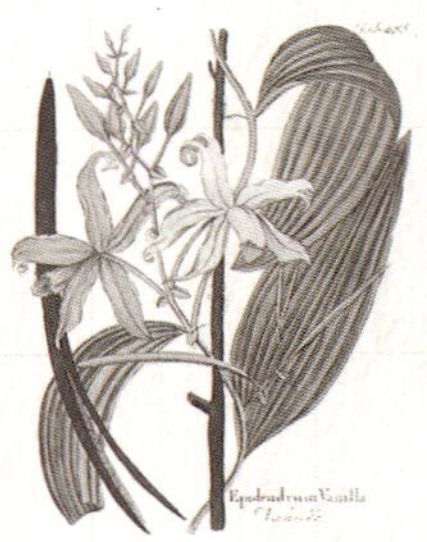

+ Scientific Name - *Vanilla planifolia*
Colonized Region Found - Madagascar
Reason for Collection - Spice

+ Scientific Name - *Coffea arabica*
Colonized Region Found - Southeast Asia
Reason for Collection - Drink

+ Scientific Name - *Helianthus annuus*
Colonized Region Found - North America
Reason for Collection - Oil & Ornamental Use

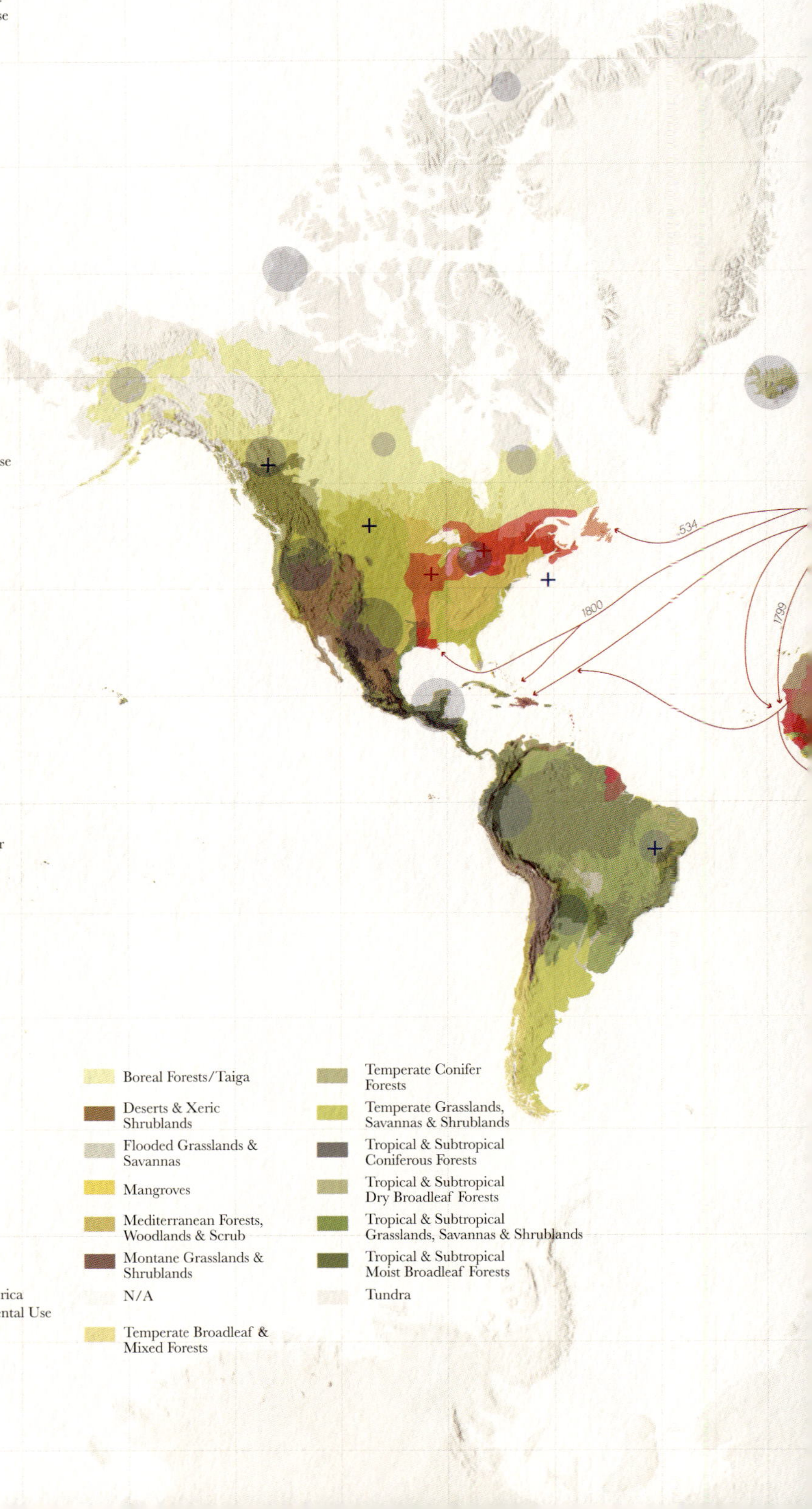

+ Scientific Name - *Clarkia franciscana*
Biome Found - Temperate Grassland, Savannas & Shrublands
Conservation Status- Endangered

+ Scientific Name - *Tabebuia aurea*
Biome Found - Tropical Savannas
Conservation Status - Secure

+ Scientific Name - *Pinus albicaulis*
Biome Found - Temperate Grassland, Savannas & Shrublands
Conservation Status - Endangered

+ Scientific Name - *Pseudophilotes sinaicus*
Biome Found - Desert & Xeric Shrubland
Conservation Status - Critically Endangered

+ Scientific Name - *Gymnoderma insulare*
Biome Found - Temperate Broadleaf & Mixed Forests
Conservation Status - Endangered

+ Scientific Name - *Senegalia catechu*
Biome Found - Tropical & Subtropical Dry Broadleaf Forests
Conservation Status - Least Concern

+ Scientific Name - *Zostera marina*
Biome Found - Intertidal Zone
Conservation Status - Least Concern

+ Scientific Name - *Thesium ebracteatum*
Biome Found - Central European Mixed Forests
Conservation Status - Least Concern

MICHELLE CHAN SYL YENG
LILLIAN CHUNG KWAN YU
KAI ZHAO ZI CHENG

GARDEN OF ANTHROPOGENIC WONDERS

The perception of exoticism is in constant flux based on power dynamics and market trends. The thirst for unique experiences, products, and aesthetics often fuels exoticism. Discarded remnants, a byproduct of our consumer-centric culture, have left a lasting impact on our ecosystems, giving rise to a new form of exoticism marked by nature's adaptability, such as the troubling discovery of a "plastiglomerate" in Hawai'i.

Setting up an experimental ground for exoticism in the Anthropocene, this scheme employs a series of walls composed of algae tiles. Daily waste is collected, classified, and processed to feed into "bio-incubator" panels. Over time and under the influence of sunlight, these tiles gradually break down various remnants and release solutions into the ponds, nurturing the formation of a novel micro-ecosystem. As visitors venture through these algae-laden walls, they step into the dynamic realm of this micro-ecosystem, marked by a spectrum of green hues reflecting the diverse stages of algae development. It is a captivating journey that underscores the intricate interplay of nature's resilience, human waste, and the evolution of life in our modern era. Foregrounding global issues instead of specificity and uniqueness, the plaza is reinvented as a symbol for the Anthropocene; a new world blooms from the obsolete, forming a common ground to remind us of our collective responsibilities regarding waste and to reflect upon our relationship to our environment.

3

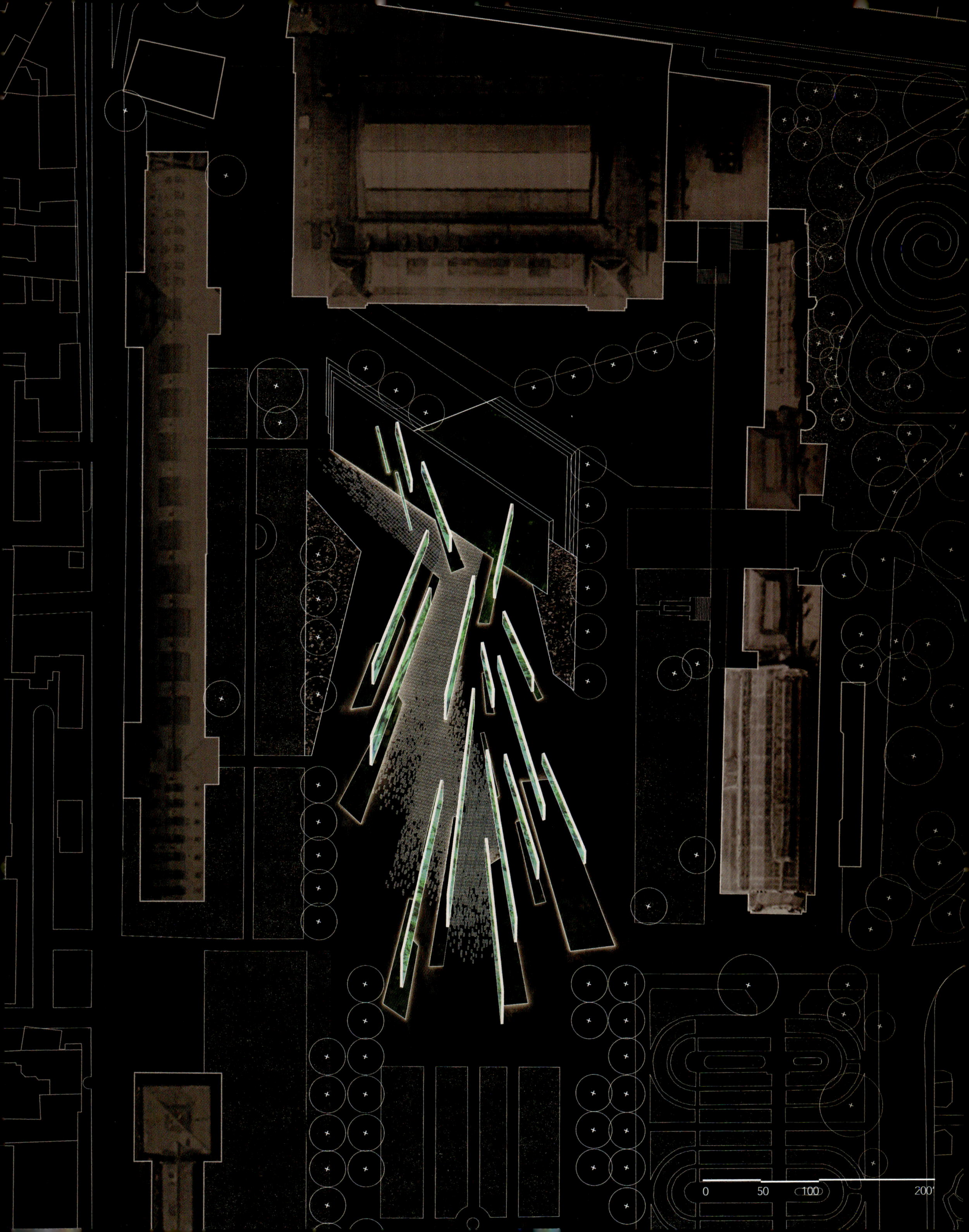
0
50
100
200'

Algae liquid flows through translucent tiles, creating different green hues during the process of digesting waste particles.

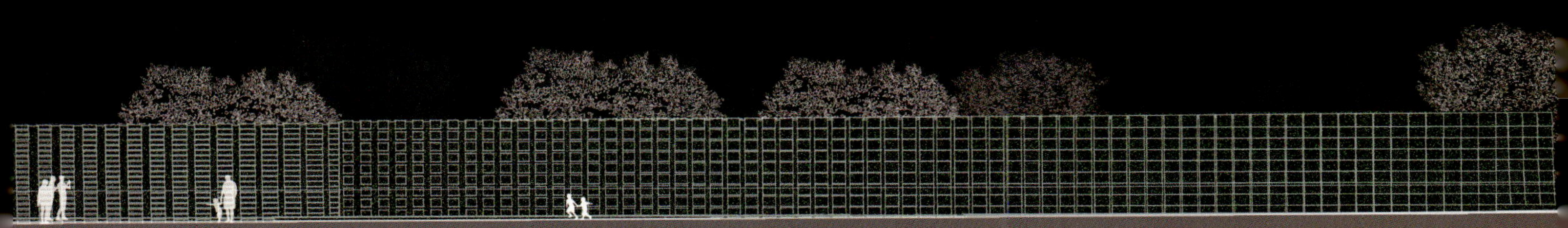

Aspergillus tubingensis
Pseudomonas sp
Amycolatopsis sp.

Ideonella sakaiensis
Bacillus sp.
Several remarkable microorganisms have evolved the ability to break down various plastic materials. While their potential for practical use is still being explored, they hold promise as natural solutions in the ongoing battle against pollution.

ISABEL YIDONG LI
OLIVER ZIYUAN ZHU

CRUMBLING EARTH
OUR FUTURE AS EXOTICISM

Exoticism can emerge from within. When we neglect the ground upon which we stand and live, and disregard the damage we inflict upon our planet, are we, the offspring of Mother Earth, the most peculiar creatures she has borne? Our actions are transforming the world into uncharted territory, where the familiar becomes strange and once-habitable places become uninhabitable. The future of our earth can be alien to us. Amid the climate crisis and the ongoing sixth mass extinction, 2023 witnessed numerous Sahara sandstorms veiling Europe, reaching as far as Scandinavia. At the same time, southern Europe grapples with desertification caused by prolonged droughts and water shortages. Reflecting on these events, our project envisions an expressive desolate landscape that serves as a memorial to extinct species.

This project seeks to transport visitors into a grim future, where the lush greenery of Jardin des Plantes turns into a barren terrain. Given the current environmental trajectory, this proposition is an increasingly plausible reality. Crumbling Earth serves as a jarring reminder of the consequences of our actions. At the heart of this landscape, the ground fractures into deep pathways, revealing the names of species driven to extinction by human activities. Visitors wander through the terrain, contemplating these extinctions as if unearthing paleontological remains. The pathway also creates a new grand entrance to the Muséum national d'Histoire naturelle, granting direct access to the Grande Galerie de l'Évolution. On top of the fractured ground, multiple stainless-steel surfaces generate dazzling visual effects meant to evoke deep emotional resonance. As this journey concludes, visitors are reunited with the beloved scenery of Jardin des Plantes and are reminded that the world can remain a place we love if we commence our efforts now.

4

In Memory of Species Made Extinct by Human Activites

Arabian ostrich
Ascension crake
Atlas wild ass
Aurochs
Atlas bear
Big-eared hopping mouse
Bluebuck
Bramble Cay melomys
Broad-billed parrot
Bubal hartebeest
Bulldog rat
Bushwren
California grizzly bear
Canary Islands oystercatcher
Cape lion
Caribbean monk seal
Carolina parakeet
Carpathian wisent
Caucasian wisent
Cebu warty pig
Chadwick Beach cotton mouse
Chatham bellbird
Chatham fernbird
Colombian grebe
Colpocephalum californici
Cuban macaw
Delalande's coua
Dodo
Domed Mauritius giant tortoise
Domed Rodrigues giant tortoise
Dusky seaside sparrow
Eastern elk
Eiao monarch
Epioblasma haysiana
Erica pyramidalis
Eudyptes warhami
Felicola isidori
Formosan clouded leopard
Goff's pocket gopher
Gravenche
Great auk
Guam flying fox
Gull Island vole
Haast's eagle
Hanyusuchus
Heath hen
Hemigrapsus estellinensis
Huia
Japanese otter
Japanese sea lion
Kangaroo Island emu
King Island emu
Laughing owl
Lesser moa
Lyall's wren
Macquarie parakeet
Madeiran scops owl
Malagasy crowned eagle
Mangareva reed warbler
Martinique amazon
Mauritius blue pigeon
Mauritius scops owl
Merriam's elk
Moa
Mount Glorious day frog
Nesoryzomys darwini
Nesoryzomys indefessus
New Zealand greater short-tailed bat
New Zealand musk duck
New Zealand owlet-nightjar
New Zealand quail
Newton's parakeet
Norfolk kākā
Noronhomys
North African elephant
North Island giant moa
North Island snipe
O'ahu nukupu'u
O'ahu 'ō'ō
O'ahu 'akialoa
Oryzomys antillarum
Pantanodon madagascariensis
Paschalococos
Passenger pigeon
Piopio (bird)
Quagga
Ratas Island lizard
Réunion giant tortoise
Robust crow
Rocky Mountain locust
Round Island burrowing boa
St. Croix macaw
San Marcos gambusia
San Martín Island woodrat
São Miguel scops owl
Schomburgk's deer
Sea mink
Small Mauritian flying fox
South Island snipe
South Island stout-legged wren
Southern black rhinoceros
Holocene extinction
Steller's sea cow
Syncaris pasadenae
Syrian elephant
Syrian wild ass
Tarpan
Tasmanian emu
Tecopa pupfish
Thicktail chub
Thismia americana
Javan tiger
Toolache wallaby
Tristramella intermedia
Ungava brown bear
Upland moa
Wake Island rail
Western black rhinoceros
Western Lewin's rail
Xerces blue

CRUMBLING EARTH
OUR FURTURE AS EXOTICISM

LANDSCAPE AS A MEMORIAL TO EXTINCT SPECIES

AN EMOTIONAL JOURNEY

Chapter 01
History of Evolution

The journey starts from the central hall of Grande Galerie de l'evolution, where a parade of animals is displayed. This reminds visitors of Earth's great biodiversity even before the existence of humans.

Chapter 02
The Fragmented Present

In the new foyer of the museum, a globe features visual representations of our present-day climate challenges, such as climate change, desertification and water scarcity. It prompts visitors to reflect on the ongoing environmental crisis that our planet faces.

Chapter 03
The Current Trajec

Upon stepping outdo
visitors are greeted w
unfamiliar forms and
textures, providing a
into a potential alien
for our planet. This in
environment piques
the curiosity of visito
encouraging them to
further and explore.

Plan 1:1000

Section 1:200

Chapter 04
The Bleak Future

Upon closer inspection of the walls, visitors will notice engraved names of species that have been driven to extinction due to human activities. Visitors wander through the terrain, contemplating these species as if unearthing paleontological remains.

Chapter 05
Moment of Reflection

Upon exiting the forecourt, visitors can see the overall forms of this bizarre landscape, reflecting on what they have passed through, and what the future of our landscape might become.

Chapter 06
Cherish Our World

Returning to the beautiful landscapes of the Jardin des Plantes, visitors are reminded of the importance to cherish and protect this beautiful planet on which we live. The world can remain a place we love if we commence our efforts now.

A NEW CLIMATE LANDSCAPE

2023: Dust from the Sahara Desert is veiling Europe
Climate change expected to result in greater intensity of such events.

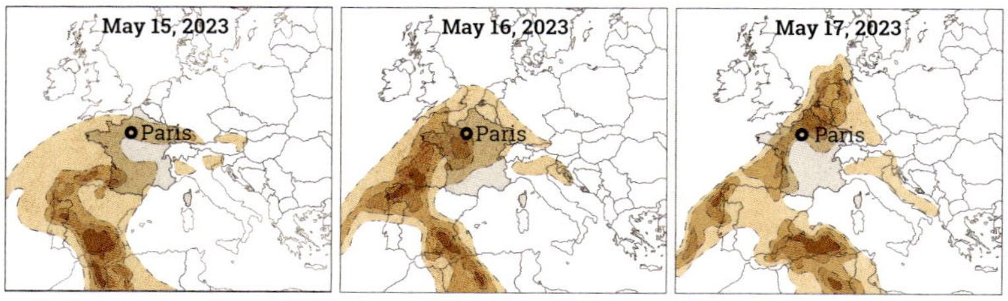

CLIMATE CHANGE

Climate change is increasing the frequency of extreme heat in Europe, which presents a threat to its residents in the years to come. Recent studies indicate that southern Europe is facing desertification due to the spread of severe droughts and water shortages throughout the continent.

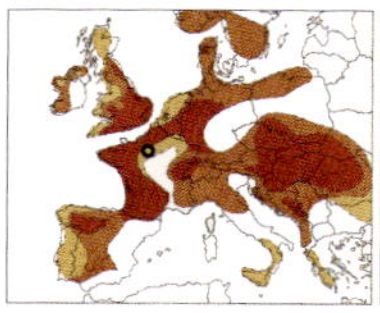

Soil moisture deficit

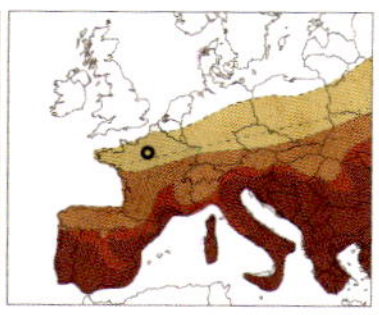

Extreme heat

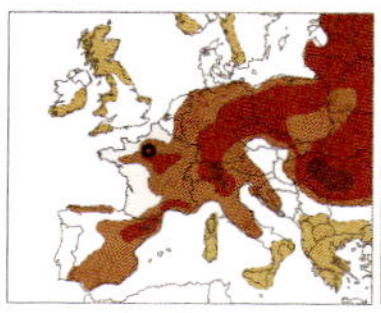

Fire danger

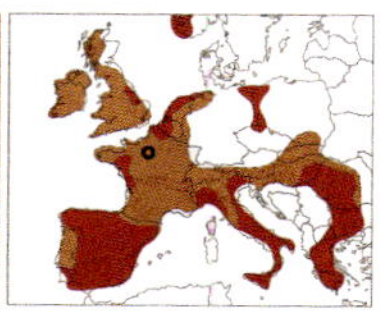

Climate Vulnerability

THE IMPORTANT ROLE OF SOIL

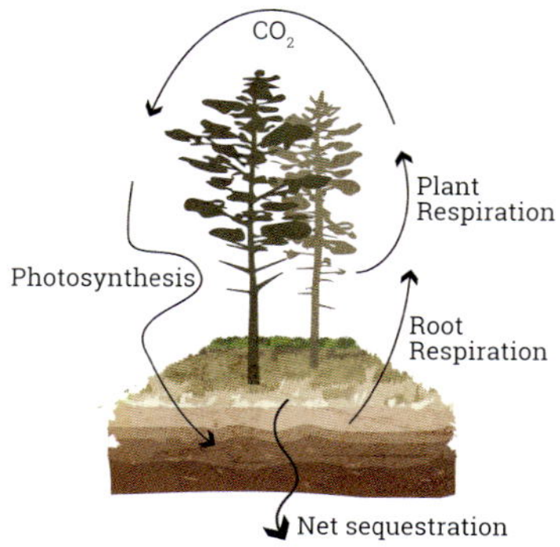

Soil plays a crucial role in maintaining ecosystem health. Around 75% of the Earth's terrestrial carbon is believed to be sequestered within the soil. When soil becomes ploughed and exposed, it becomes a significant contributor to global warming by releasing both carbon dioxide and methane.

DESIGN INSPIRATION

The Jardin des Plantes, situated in the heart of Paris, serves as a cherished tribute to the wonders of nature and its diverse offerings to us. It contains gardens, archives, libraries, greenhouses, museums and a zoo.

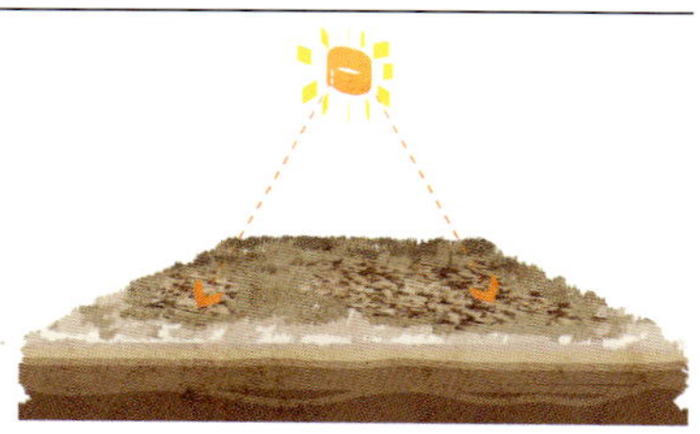

The current climate trajectory suggests a bleak outlook for the future. It is estimated that Europe will experience extreme heat and droughts, leading to the desolation of many of its cherished natural landmarks.

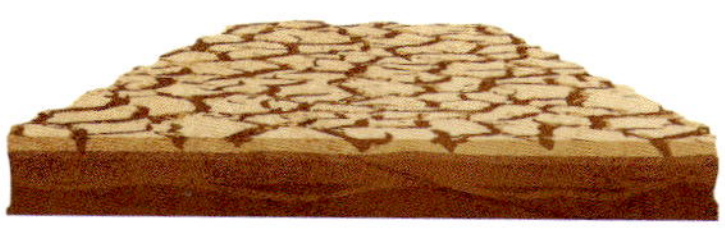

The project imagines a desolate landscape in an alternate future. It serves as a jarring reminder to all visitors, vividly illustrating the consequences of our own actions. Given the current trajectory, this proposition is not mere fantasy but an increasingly plausible reality.

The terrain fractures, unveiling a profound pathway at its core. Visitors wander through the terrain, contemplating these extinct species as if unearthing paleontological remains. Additionally, this design introduces a new entrance to the existing museum.

View - Main Crack of Fragmenting Land

View - Crumbling Surfaces

Caribbean monk seal
Carolina parakeet
Cape lion
Bush wren
HOPE IS NOT LOST...

ADRIEN ROUSSEAU

THE STONE NESTS

In the heart of bustling Paris, this bold and poetic aerial work combines the artifice of the city with the majesty of nature. Made of multiple concrete columns assembled within a light metal structure, resembling scaffolding, the "stone nests" evoke the curious yet familiar shapes of stalactites and stalagmites, termite mounds, or even tree bark, seemingly stepping out of a Jules Verne novel. The sinuous mineral columns at times rise skyward as well as plunge into the ground. They evoke the ancient art of rockwork, which brought the picturesque and exotic to 19th-century Parisian parks and gardens.

The columns are designed to welcome and observe birds. Each one features a unique combination of colors and textures. Carefully carved-out holes are integrated, providing birds with safe places to nest and raise their young. The cavities vary in configuration and size to accommodate the broadest possible range of species. Located within the main square of the Muséum national d'Histoire naturelle, this open aviary would become the new visual landmark along the axis of the Jardin des Plantes. The stone nests evolve with the habitat they support, inviting visitors to look up and marvel. This sanctuary for our winged friends reminds us that the meanders of concrete can be designed to meet the demands of living things. Every morning, the garden awakens to the chirping of birds, a music that has regained its place in the city's frenetic rhythm thanks to these avian columns.

4

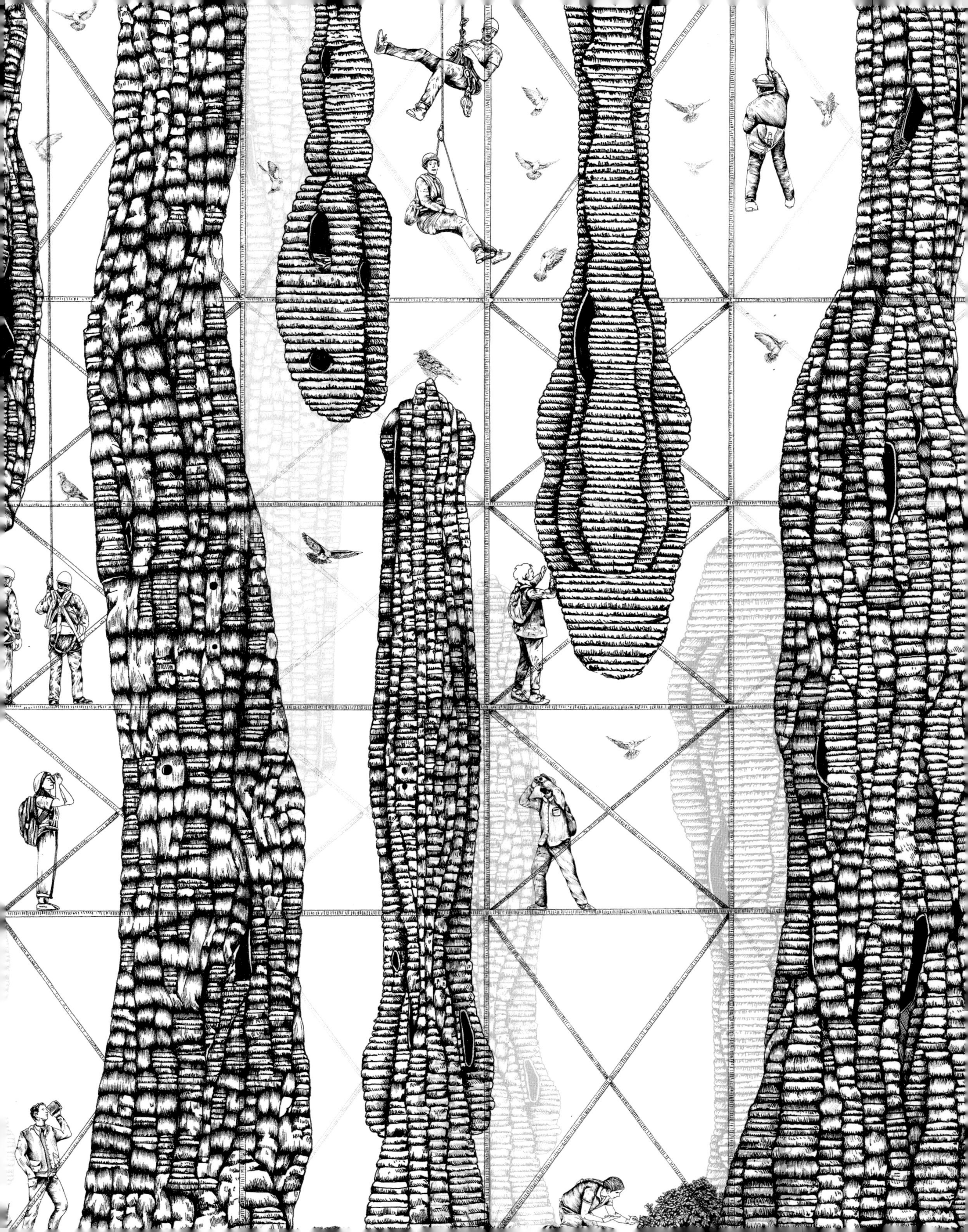

THE ART OF IMITATING NATURE IN DIFFERENT PLACES AND AT DIFFERENT TIMES IN PARIS

Buttes Chaumont Park
1867
Fake stalagmite made in concrete.

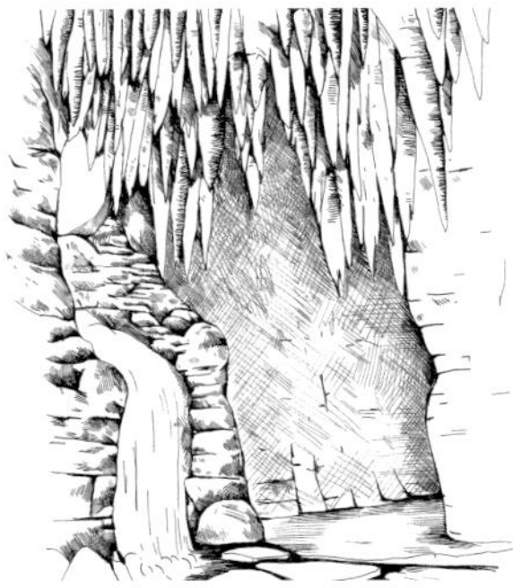

The technique of "Rusticage" was very popular during the second half of the 19th century. Thanks to advances in reinforced cement, it offered masons of the time the possibility of decorating works, such as kiosks, balustrades, benches, bridges, stairs, guardrails, ramps with imitations of branches or tree trunks modeled in fresh mortar to give them a rustic appearance.

Garden of the Eiffel Tower
1877
Rusticage - fake wood structure in concrete

This technique was also used for the rocks and fake caves in the garden of the Eiffel Tower 10 years later. It allowed the creation of a fantastical and bucolic setting for the gardens at the 1878 Universal Exhibition.

Vincennes Zoo
1934
Fake Rock in concrete

Under the sprayed concrete on a r mesh, there is a structure of reinfo concrete posts and beams. The id of an artificial rock in zoos was not exclusive to Vincennes. Inaugurate 1934, it was inspired in particular b of the Hamburg zoo, already in se in 1907.

des Plantes

nt structure

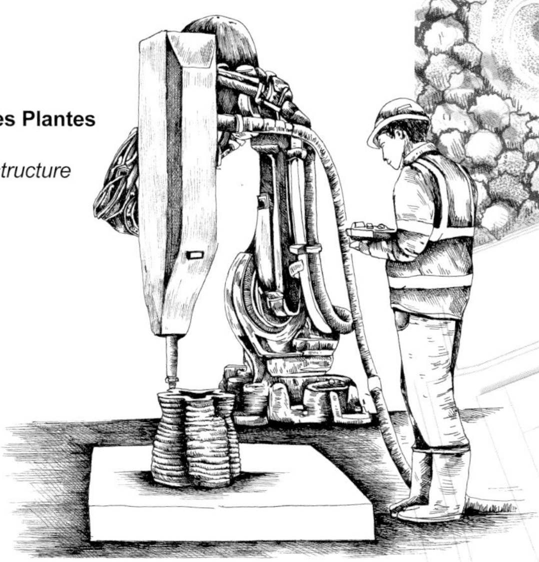

deled within the Museum by ts, the columns are reproduced in full size using large-scale ction 3D printers. Different types rete are used depending on the es, bringing a diversity of shapes, s and shades to the whole.

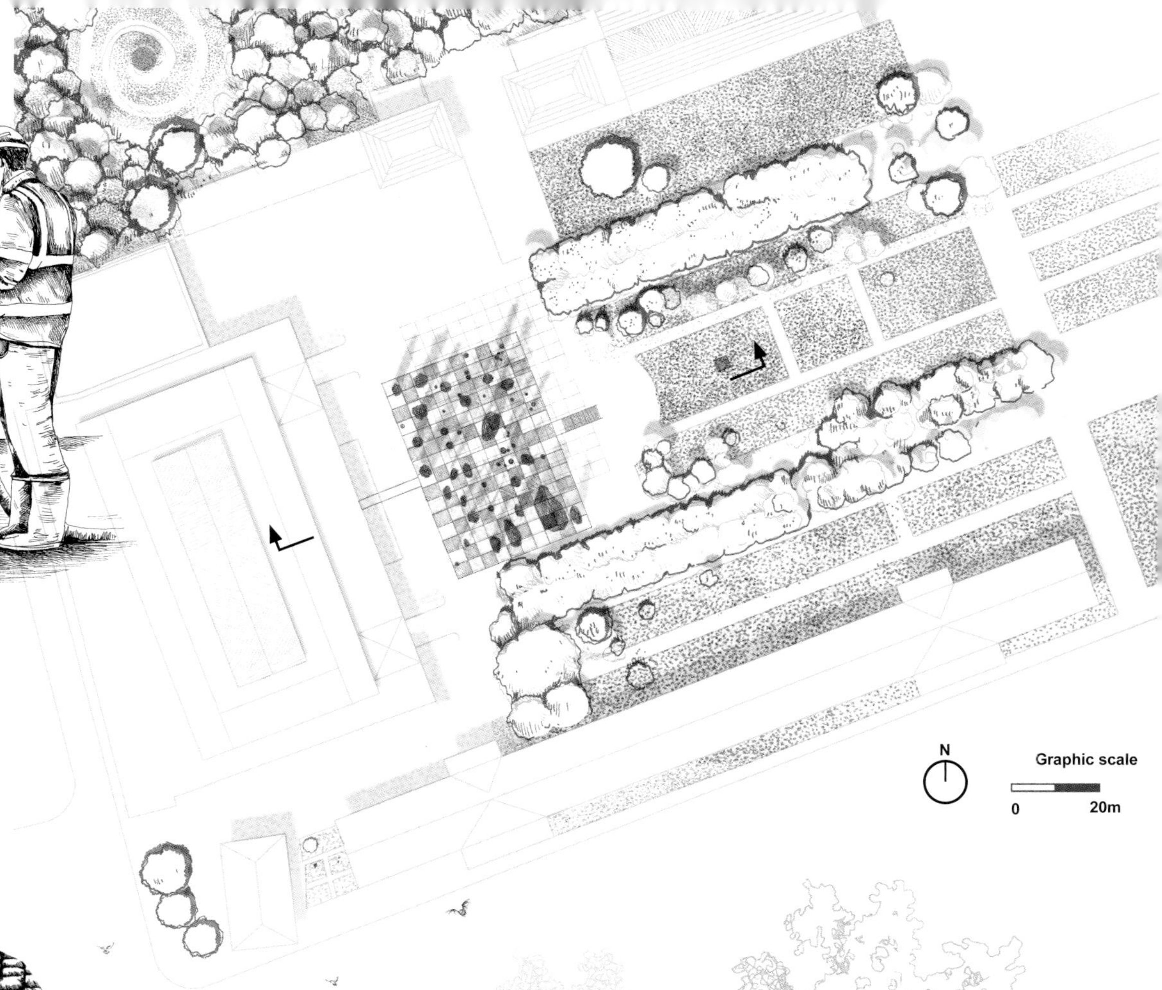

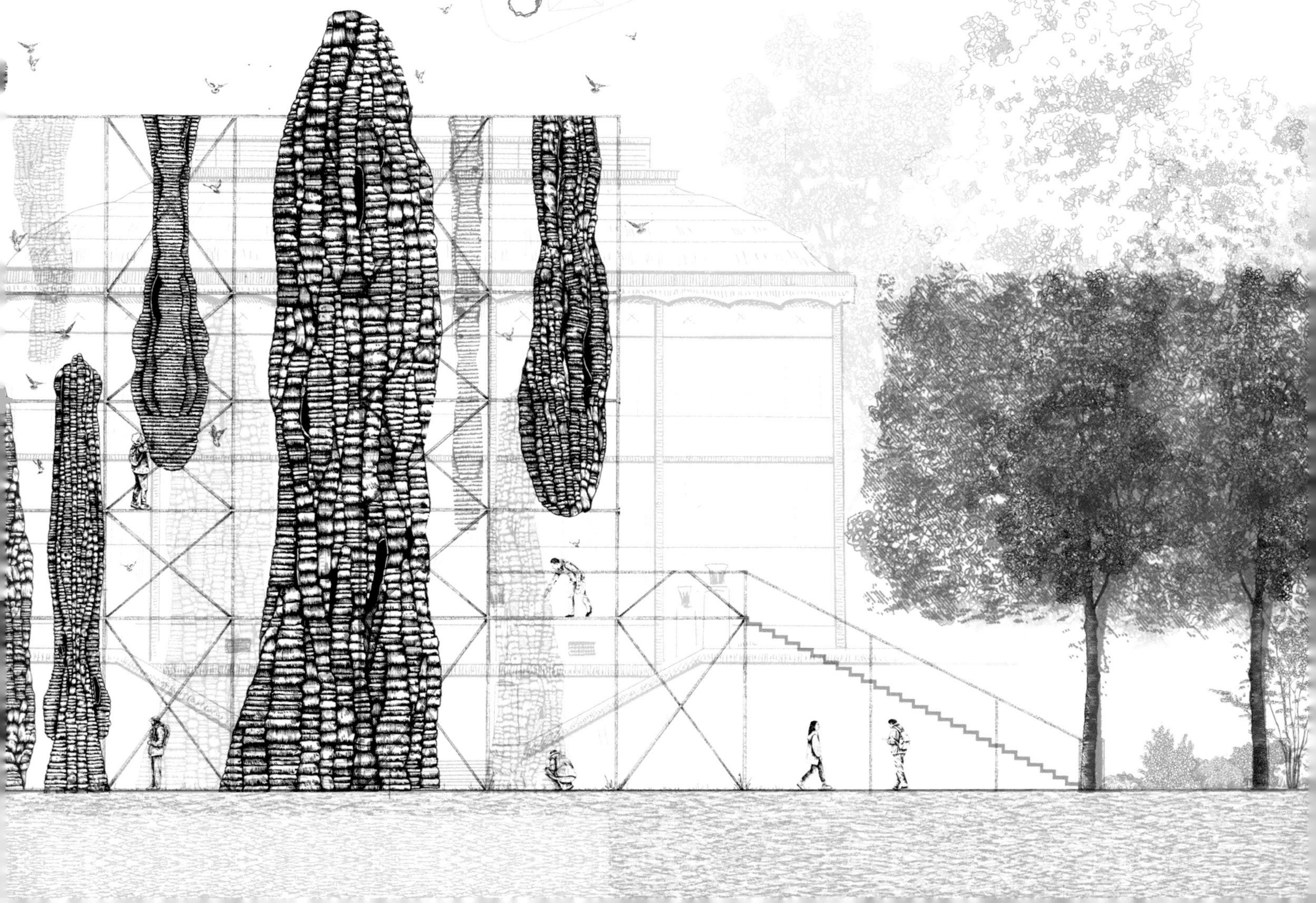

Climbing is the only met
to study the nests and li
inside the least accessib
structures.

The annual scientific sur
offer a vertical choreogr
disturbed by the flight of
different birds.

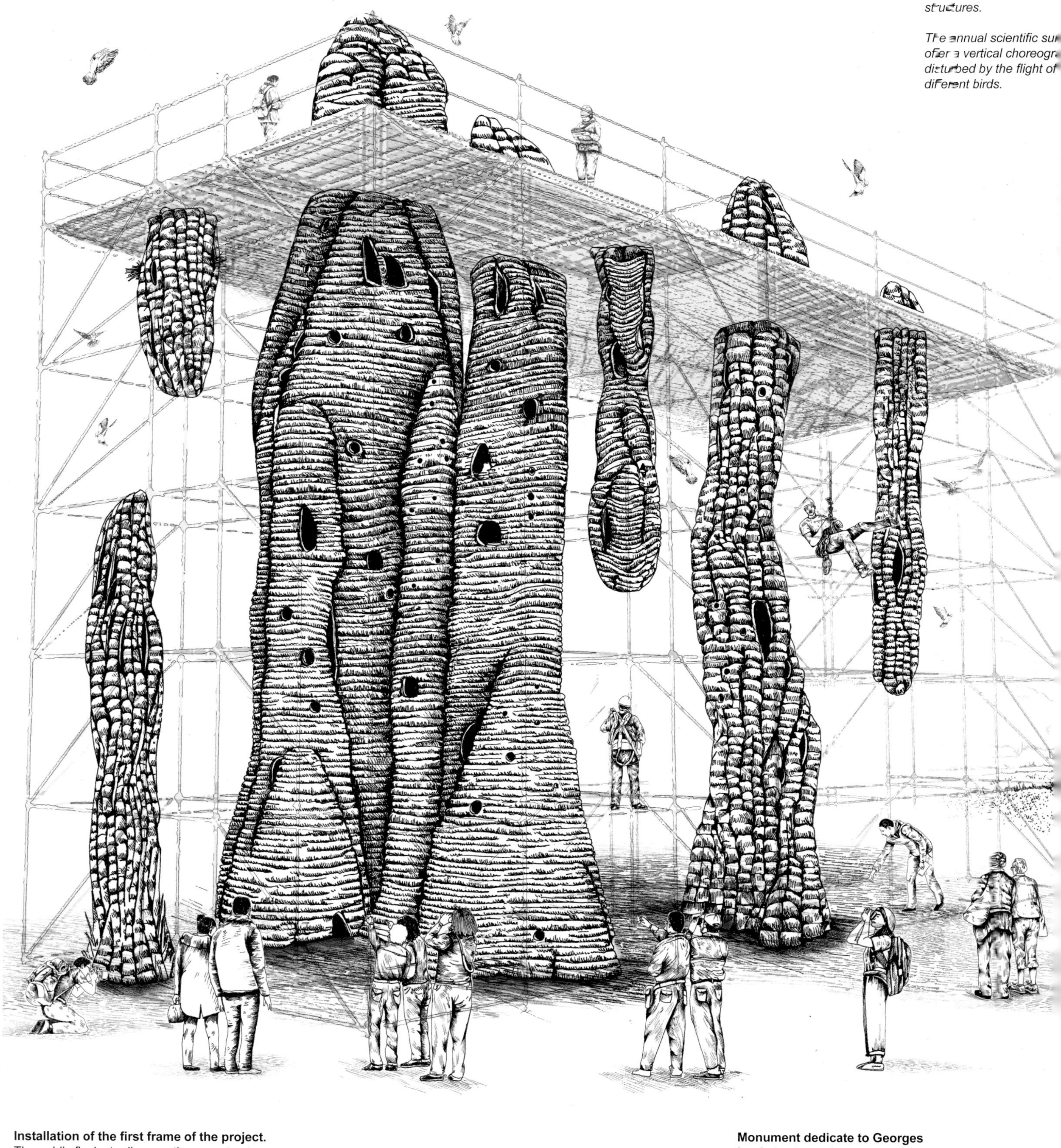

Installation of the first frame of the project.
The public flocks to discover the new structures that have just come off the 3D printers.

The challenge for scientists is to position these elements outside of nesting periods and not to disturb the birds who have already nested.

Monument dedicate to Georges Louis Leclerc - Comte de Buffon (by Jean Carlus)
1707-1788
He is represented holding a bird in his hand. Nine volumes of his encyclopedia are dedicated to birds (from 1770 to 1783).

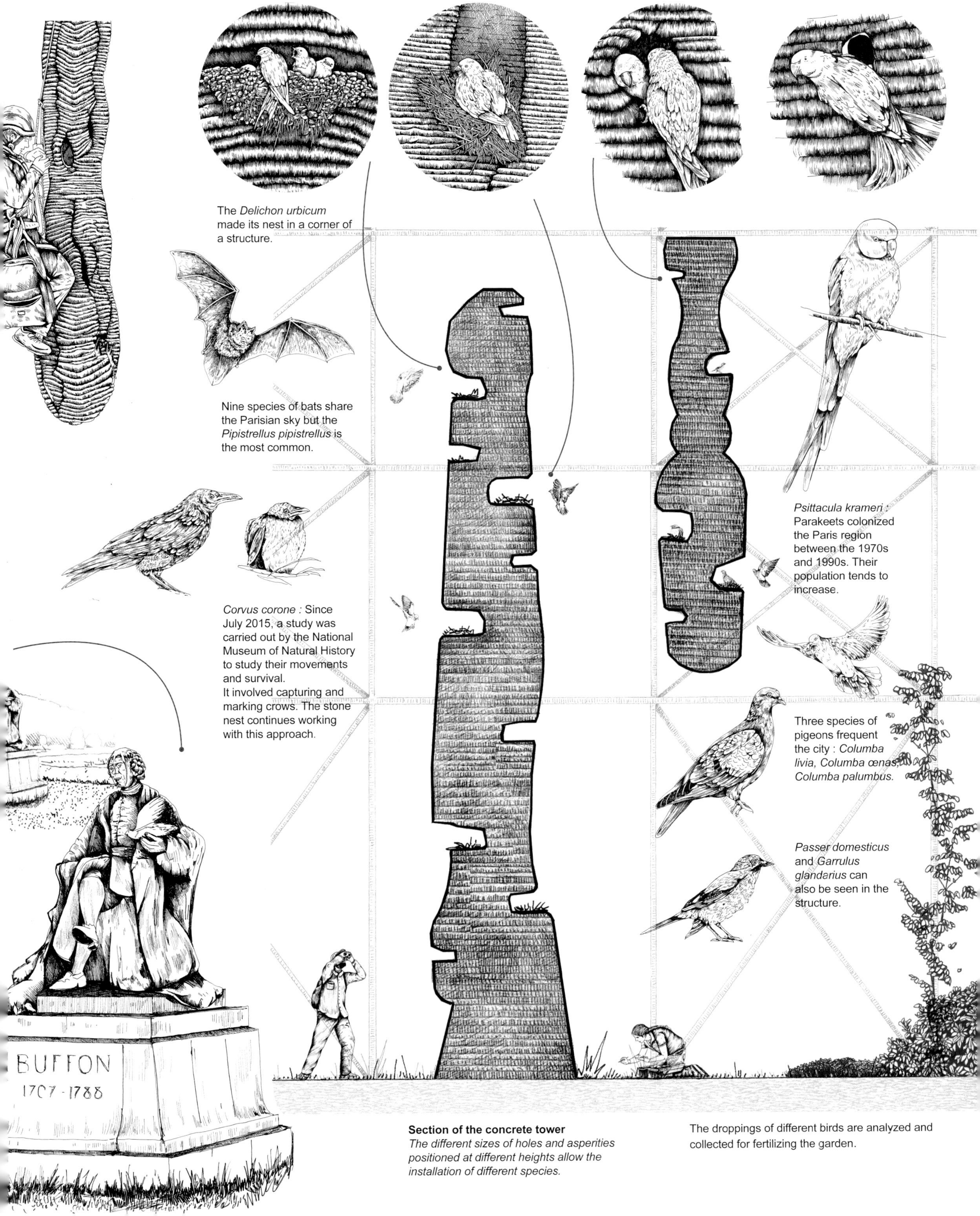

Section of the concrete tower
The different sizes of holes and asperities positioned at different heights allow the installation of different species.

The droppings of different birds are analyzed and collected for fertilizing the garden.

HONORABLE MENTIONS

THE BLACK AXIS

CHUANQI LIU • MUYUN XIAO • WENJIA ZHANG

The Black Axis is a 50-meter concrete tower situated in front of the Muséum national d'Histoire naturelle. This exotic and abstract object stands in the museum's forecourt, providing a space for visitors to reconsider their relationship with the nonhuman. As visitors approach the Axis from in front of the statue of Comte de Buffon, they step into a deep well and come face to face with an array of statues of species that have gone extinct over the past two centuries. This space implies a reversal of mental iconography, with humans being gazed upon by nonhuman creatures. As one ascends, the space becomes narrower. Upon reaching the top, visitors can see the Eiffel Tower, a symbol of technological mastery. After circling the top of the tower, visitors slowly descend the back of the well, which is engraved with messages about the extinct species. This is the last judgment of the extinct species.

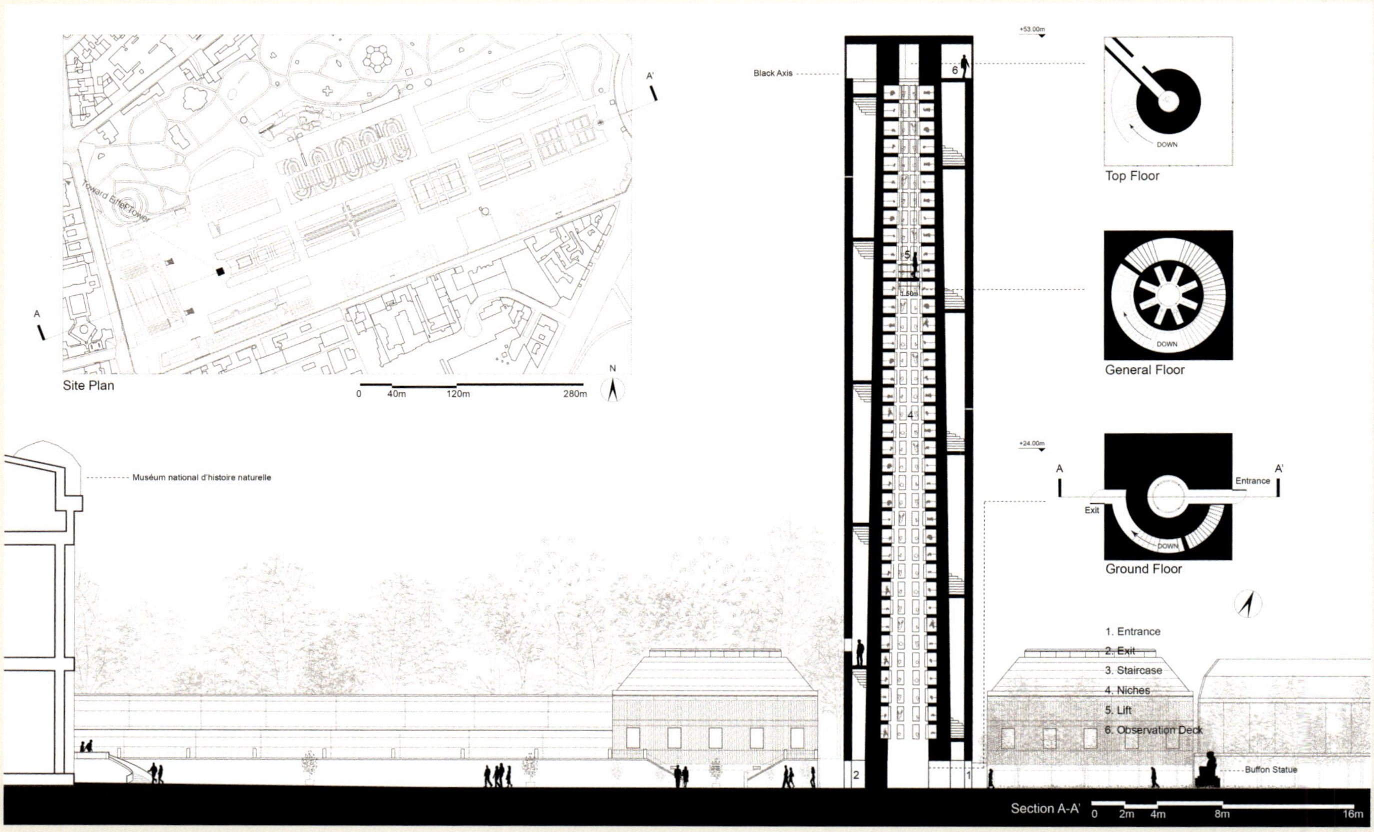
Site Plan
Toward Eiffel Tower
0 40m 120m 280m
N
A
A'
Black Axis
Muséum national d'histoire naturelle
+53.00m
+24.00m
Top Floor
General Floor
Ground Floor
DOWN
Entrance
Exit
1. Entrance
2. Exit
3. Staircase
4. Niches
5. Lift
6. Observation Deck
Buffon Statue
Section A-A'
0 2m 4m 8m 16m

1. View through niches and species statues

2. View of Eiffel Tower from lift exit on the top floor

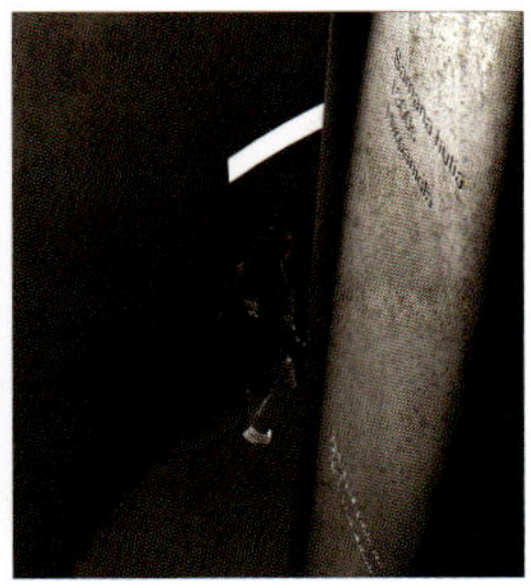

3. View of the back of the niche

A FIELD OF THE ORDINARY

YANG FEI • XINYI ZHOU

Set against a backdrop rich in exotic collections and colonial treasures, A Field of the Ordinary is a serene intervention juxtaposing an assembly of 260 trees native to France. Placed at the end of the Jardin des Plantes central allée, visitors will find an array of trees of varying heights, surrounded by red concrete planks that resemble scars on the ground. This array contrasts with the prevailing symmetry and grandeur of the site's axial layout. Visitors are expected to be guided by the flow of concrete planks as they move toward the heart of the field, where the planks get shorter and heavier and the ground staunches like a pond. This area is meant to be a sanctuary for contemplation, where visitors can sit under a dome of tree canopies.

Over time, as the trees mature and grow taller, they will gradually reshape the field, diminishing the dominant presence of the monolithic Grande Galerie de l'Évolution along the site's central axis. This subtle transformation mirrors a healing process, echoing the recovery over time from the historical damages that imperialism inflicted upon colonized individuals, communities, and cultures. Amidst the field, a mutual understanding and collective reimagining of what constitutes the "exotic" may emerge.

...penetrated deeply into the field composed of trees and concrete planks... like a river that slowly staunches.
Walking down the tree-lined Allée, an unfamiliar touch of green masking the grand gallery.

Front elevation
Plan

Painting: RECEPTION OF THE AMBASSADORS FROM SIAM AT THE CHÂTEAU DE FONTAINEBLEAU
Both the real 1861 ceremony and Gérôme's painting of it aimed to legitimize the Second French Empire and its colonial ambitions in Asia through this renewed alliance with Siam against Britain.
A field that offers a brief escape from the narrative of exoticism that is inherently Western and colonial.

THE FOREIGN INSIDE OF US

JULIA TREICHEL

The site is characterized by its rich history and strategic location next to the Seine River, as well as by the architectural context of the Jardin des Plantes and museums. The colonial structures must be scrutinized and reflected upon.

This scheme uses a vast water surface to fill the space of the forecourt, serving as a large mirror of the surroundings. It also establishes a connection with the Seine and can act as a cooling agent for the city. Through this new element, visitors are guided along narrow paths. The reflection of the environment overlaps with the reflection of the visitor on the water surface, making them a part of the exhibit and prompting them to question their role and position. Instead of showcasing "exotic" displays, visitors are invited to become the object of the exhibition. In small, "floating" pavilions, from behind iron bars, visitors can observe unusual perspectives from both outside and inside the cages, shifting hierarchies in the process. Another central design element is the expansion of the medicinal gardens, which were the original gardens planted at this site. The pavilions are designed to create intense aromatic atmospheres using a diversity of medicinal plants grown in beds around the water basin and in the pavilions. Visitors would learn about the plants' effects and even participate in their cultivation.

THE FOREIGN INSIDE OF US
A Landscape Approach to Reflect
on "THE EXOTIQUE, THE OTHER AND THE SELF"

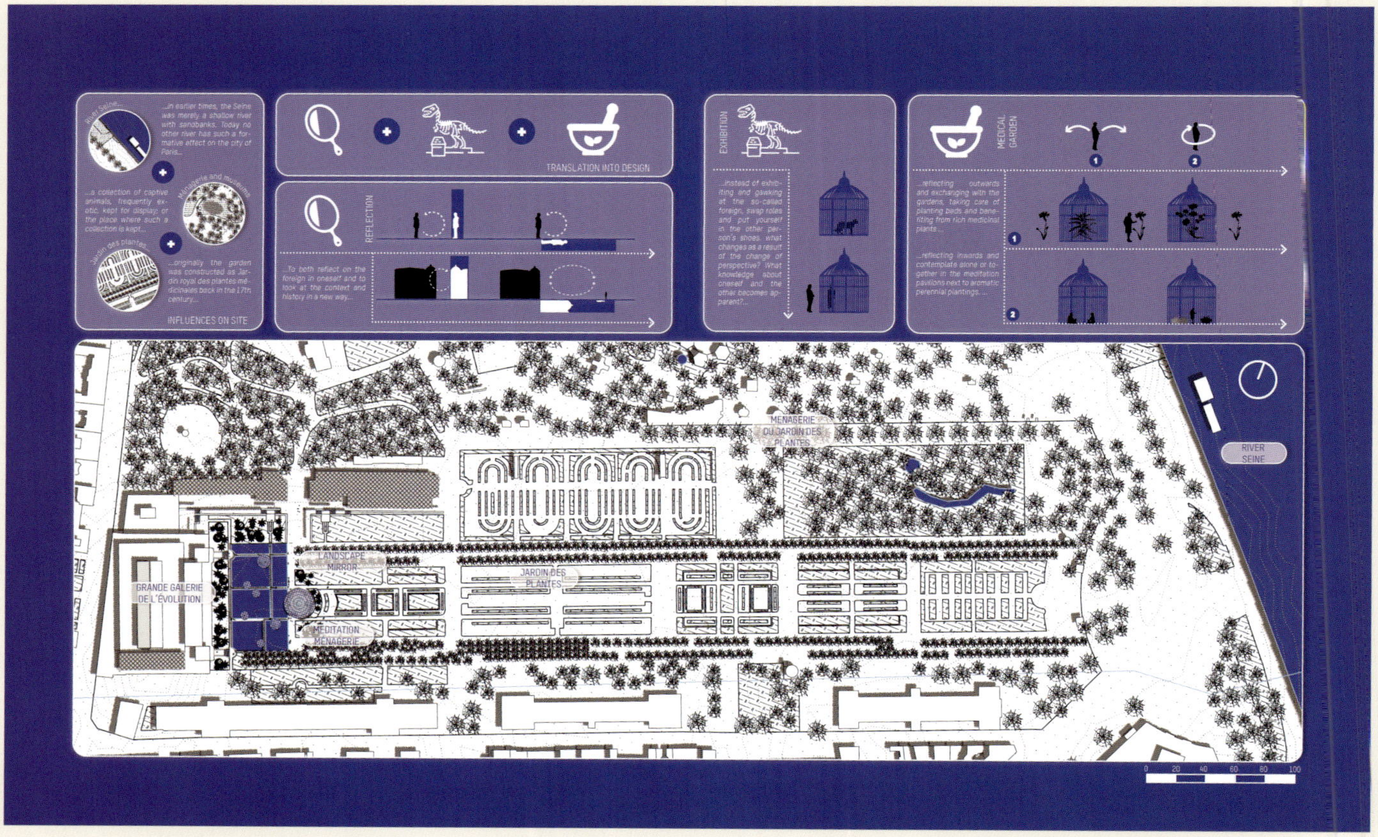
River Seine...
...in earlier times, the Seine was merely a shallow river with sandbanks. Today no other river has such a formative effect on the city of Paris...
...a collection of captive animals, frequently exotic, kept for display; or the place where such a collection is kept...
Menagerie and museums
Jardin des plantes...
...originally the garden was constructed as Jardin royal des plantes médicinales back in the 17th century...
INFLUENCES ON SITE
TRANSLATION INTO DESIGN
REFLECTION
...To both reflect on the foreign in oneself and to look at the context and history in a new way...
EXHIBITION
...instead of exhibiting and gawking at the so-called foreign, swap roles and put yourself in the other person's shoes. what changes as a result of the change of perspective? What knowledge about oneself and the other becomes apparent?...
MEDICAL GARDEN
...reflecting outwards and exchanging with the gardens, taking care of planting beds and benefiting from rich medicinal plants...
...reflecting inwards and contemplate alone or together in the meditation pavilions next to aromatic perennial plantings...
MENAGERIE DU JARDIN DES PLANTES
RIVER SEINE
LANDSCAPE MIRROR
GRANDE GALERIE DE L'EVOLUTION
JARDIN DES PLANTES
MEDITATION MENAGERIE
0 20 40 60 80 100

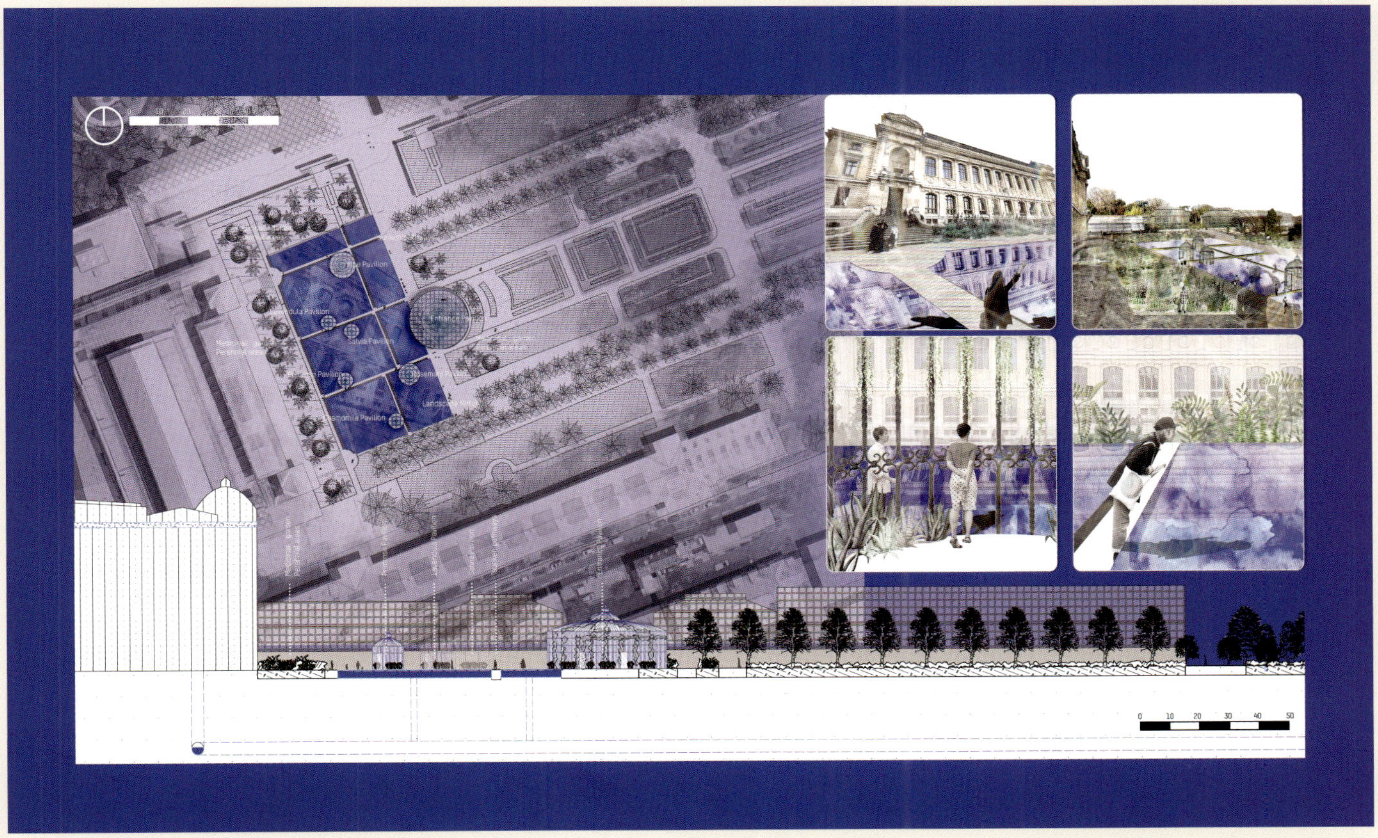

GINKGONOMY

FARZIN BAIK

Approaching the museum's grand forecourt, one is met with the imposing façade of the Grande Galerie de L'Évolution, symbolizing human curiosity and our quest to understand Earth's living tapestry. This forecourt is a symbolic gateway to the world of taxonomy, bridging past and present, tradition and innovation, subjectivity and objectivity. Many of the museum's collections are categorized and labeled based on Western European perspectives, which often results in the underrepresentation or misrepresentation of non-European cultures, species, or ecosystems. This can promote colonial narratives and impede a more inclusive understanding of natural history.

This proposal aims to expose and address the limitations of the historical bias in taxonomy. It emphasizes the dynamic nature of evolution and creates sensory immersion through dense plantings of many ginkgo cultivars. *Ginkgo biloba* presents a challenge to traditional taxonomic boundaries. It is a "living fossil" that has remained relatively unchanged for 200 million years and is the only living species in the genus. This raises questions about how we classify species that occupy unique evolutionary niches. To celebrate the ginkgo tree and enhance the sensory stimuli of the forecourt, Ginkgonomy places them in misty planting beds and combines them with artificial elements, such as imbricated super-leaves, wind chimes, and wind spinners inspired by the shapes of ginkgo leaves.

Hanging Aluminum Wind Chimes
Short Aluminum Wind Spinners
Tall Aluminum Wind Spinners
Artificial Imbricated Super-leaves
Great Gallery of Evolution
Elev. +128
Large Ginkgo Planter
Barabits fastigiata ginkgo
Chase Manhattan ginkgo
Autumn gold ginkgo
Santa Cruz ginkgo
Elev. +130
Small/Medium Ginkgo Planters
Beijing gold ginkgo
Mariken dwarf ginkgo
Pendula ginkgo
Void/ 2' Deep Foggy Pond below
Wood Deck
Large Ginkgo Planter
Barabits fastigiata ginkgo
Chase Manhattan ginkgo
Autumn gold ginkgo
Santa Cruz ginkgo
Sec. B
Elev. +118
Sec. A
Terraced Foggy Pond
Underpass Gallery
Elev. +131
Elev. +117
Elev. +116
Artificial Imbricated Super-leaves
Artificial Imbricated Super-leaves in the Foggy Pond
Large Ginkgo Planter
Barabits fastigiata ginkgo
Chase Manhattan ginkgo
Autumn gold ginkgo
Santa Cruz ginkgo
Sec A.
0 15' 30'

Ginkgo biloba:
• is an ancient survivor;
• is an air quality indicator;
• has unique genetic features;
• is hardy in urban landscapes;
• has slow molecular changes;
• is pest and disease resistant;
• has its own unique branch: Ginkgophyta;
• coevolved with now-extinct animals;
• has vibrant golden-yellow fall foliage;
• is widely planted for its aesthetic beauty;
• potentially improves memory and cognitive health;
• has edible seeds that are used in Asian cuisine;
• is a symbol of longevity, endurance, and duality In Chinese culture;
• has fan-shaped leaves split into two lobes, resembling the shape of a duck's foot.
Great Gallery of Evolution
Wooden Deck Underpass Pond-side
2' Deep Foggy Pond
Artificial Imbricated Super-leaves
Wooden Ramp
Wind Catcher
Beijing gold ginkgo
Mariken dawrf ginkgo
Underpass Gallery
Autumn gold ginkgo
Santa Cruz ginkgo
Sec B.
0
15'
30'

GARDEN OF EXTINCTION

MAURA MCDANIEL • ISAIAH SCHAREN

We exist in a time of widespread extinction and unprecedented changes in the climate, undeniably driven by human actions. As we collectively scramble to mitigate these devastating effects, conservationists are attempting to prevent species loss through *in situ* and *ex situ* work. This is a speculative proposal to "conserve" *ex situ* a piece of back mangrove from Mayotte, a small archipelago of islands approximately 200 miles off the northwest coast of Madagascar. According to the Red List of ecosystems in France, this landscape is at risk of destruction by embankment and urbanization. We chose this area based on its abundance of endemic species and at-risk ecosystems and its status as a French overseas department. In an extreme measure of preservation work, we propose to relocate 2,500 m2 of plant and animal material from Mayotte to the museum, destroying the site of origin in the process. The specimens will become bronzed artifacts in an installation in the museum's forecourt. The installation will be a one-to-one replication of the "conserved" land area and accompanying specimen samples.

Large institutional collections uproot locality by collecting samples from around the globe and across time, reassembling them into a constructed context. This proposal is not an outright critique of the museum or of larger conservation practice. For us, it is a reflection on a deep internal conflict between the desire to see and know the world and a profound sadness that, in the not-so-distant future, the only opportunity to do so may be through hollow artifacts of a world we lost.

GROWTH HABIT
PHYLUM
CLASS
ANTHROPODA
REPTILIA
AVES
Wild date palm
Seashore dropseed
Sporobolus virginicus
Golden fern
Acrostichum aureum
Hibiscus tiliaceus
Broad Lipped Acampe
Acampe pachyglossa
Copepods
Copepoda
Fiddler Crab
Tubuca urvillei
Red mangrove crab
Neosarmatium meinerti
Cardisoma carnifex
N/A
Macrophtalmus depressus
N/A
Sesarmops impressus
Mangrove Oyster
Atlantic mudskipper
Dorsal-lined day gecko
Mayotte White-eye
Zosterops mayottensis
Painted Dove
Comoros Founingo

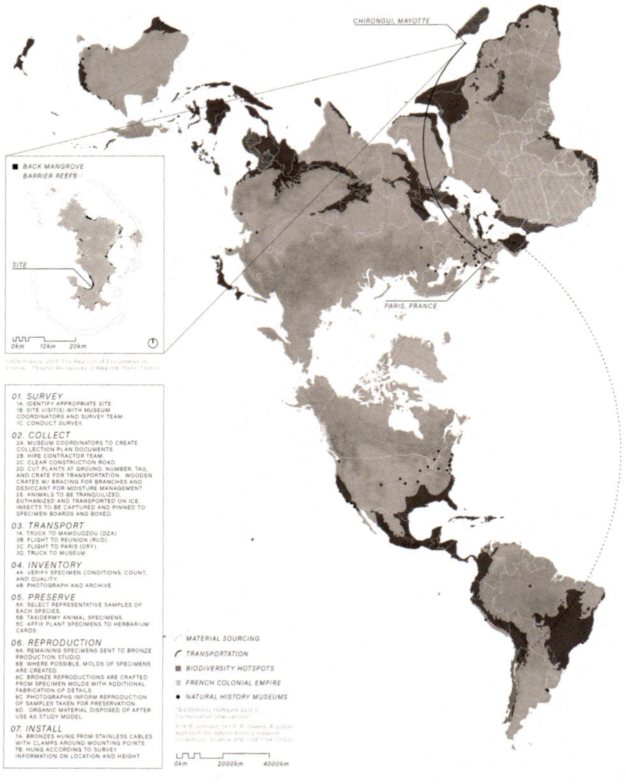

SITE OF ORIGIN *CHIRONGUI, MAYOTTE*

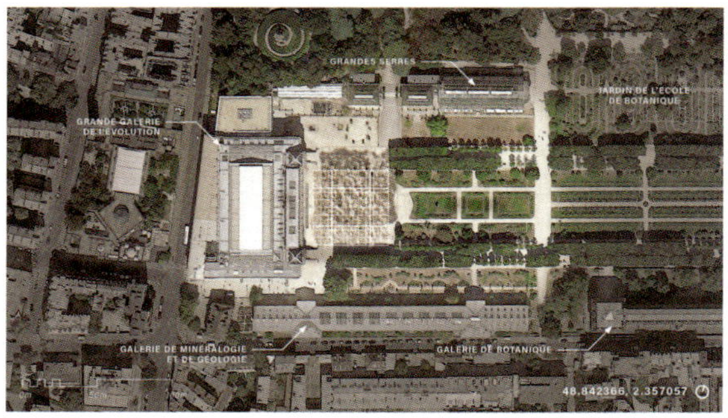

SITE OF PRESERVATION *PARIS, FRANCE*

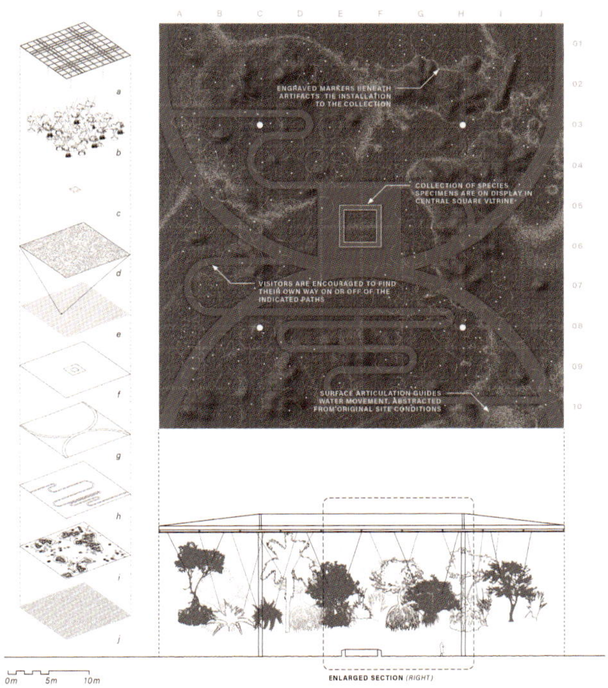

Measured point elevations, observed litter, and digital surface scans from site survey produce an abstracted ground plane that is translated into the 1m x 1m granite tile of the plaza. Smooth paths encourage ambulation while engraved markers encourage exploration.

- *a* **SUPERSTRUCTURE**
- *b* **SUSPENDED ARTIFACTS**
- *c* **SPECIMEN DISPLAY CASE**
- *d* **SAMPLED SURFACE TEXTURE**
- *e* **1m x 1m TILE SUBDIVISION**
- *f* **COLLECTION CIRCULATION**
- *g* **PRIMARY PATHS**
- *h* **SECONDARY PATHS**
- *i* **OBSERVED LITTER DISTORTION**
- *j* **1m x 1m POINT ELEVATIONS**

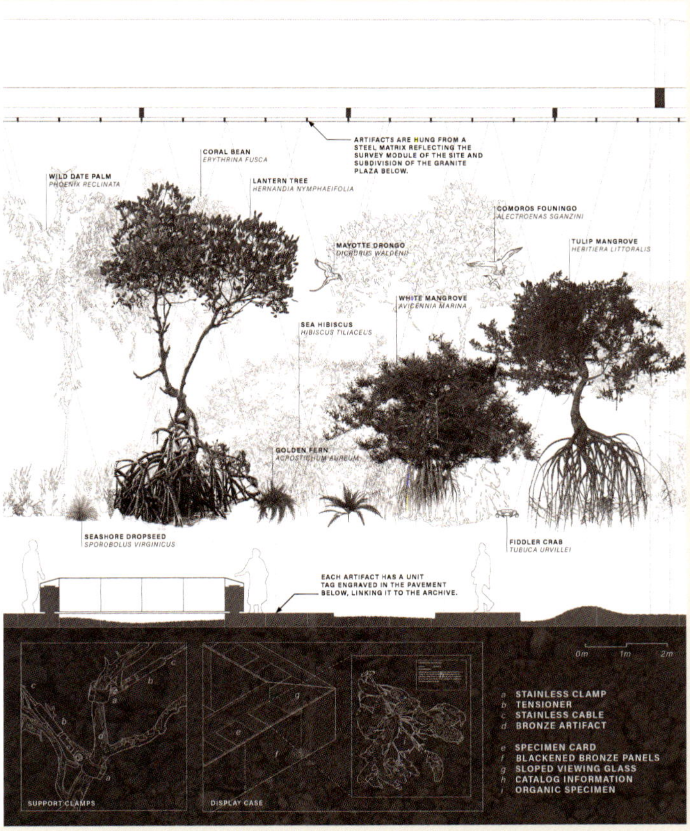

THE DECLASSIFYING FORECOURT

DANIEL COOMBES

The forecourt of the Muséum national d'Histoire naturelle is surrounded by collections of vegetal, mineral, and animal origin. Soil is a material common to all the collections and simultaneously missing from this museum landscape. While the Galerie de Géologie et de Minéralogie holds "Martian meteorites, giant crystals, rocks, and admirable minerals," there is no information about soils. Are they missing because of their ubiquity? Or because the presence of land from around the world problematizes the practices of natural history collections and amplifies what the existing collections and their displays enact?

The Paris Herbarium database is used as the basis of this proposal. The herbarium was started in 1650 and contains approximately six million specimens from 247 countries, with 10,000 specimens added annually. The design strategy stems from a simple equation: one liter of soil for each specimen within the collection is delivered to the site. Thus, the Declassifying Forecourt accommodates 6,000 cubic meters of soil, aligning with the six million specimens in the Galerie de Botanique's herbarium. Like all material that enters this landscape, the soil must be classified within a universal system. The current soil taxonomy is the World Reference Base for Soil Resources, which includes 32 types. The soil is further divided into eight groups based on shared characteristics. In this design, which links the quantity of soil to the herbarium collection, the sheer mass of the herbarium is revealed. The countries most represented in the herbarium–i.e., France and its current and former colonies–form the largest pile of soil; thus, while the soil is configured by soil taxonomy as it is placed in the forecourt, the uneven quantities suggest geographic and colonial influences.

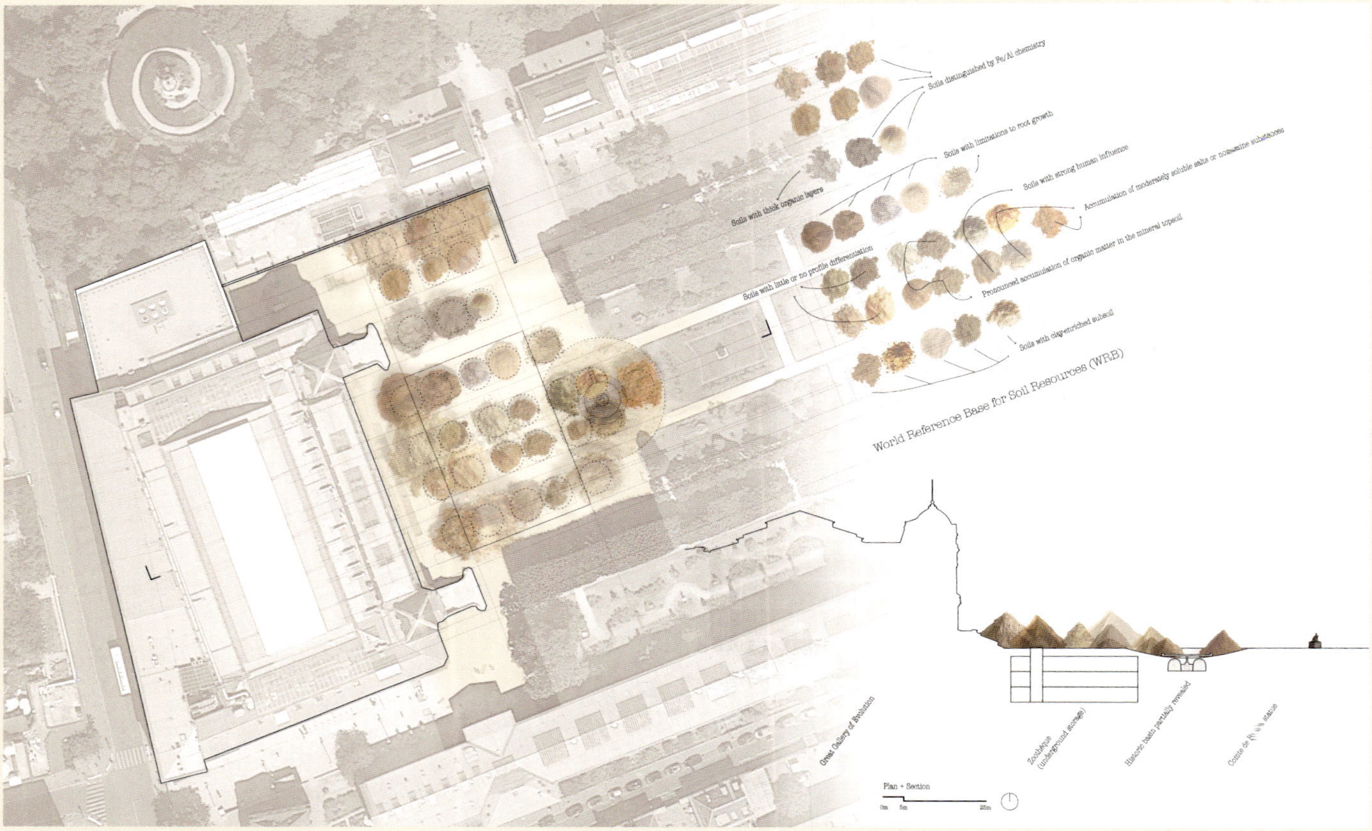
Soils distinguished by Fe/Al chemistry
Soils with limitations to root growth
Soils with thick organic layers
Soils with strong human influence
Accumulation of moderately soluble salts or non-saline substances
Soils with little or no profile differentiation
Pronounced accumulation of organic matter in the mineral topsoil
Soils with clay-enriched subsoil
World Reference Base for Soil Resources (WRB)
Great Gallery of Evolution
Zoothèque
(underground storage)
Historic basin partially revealed
Plan + Section
0m 5m 25m

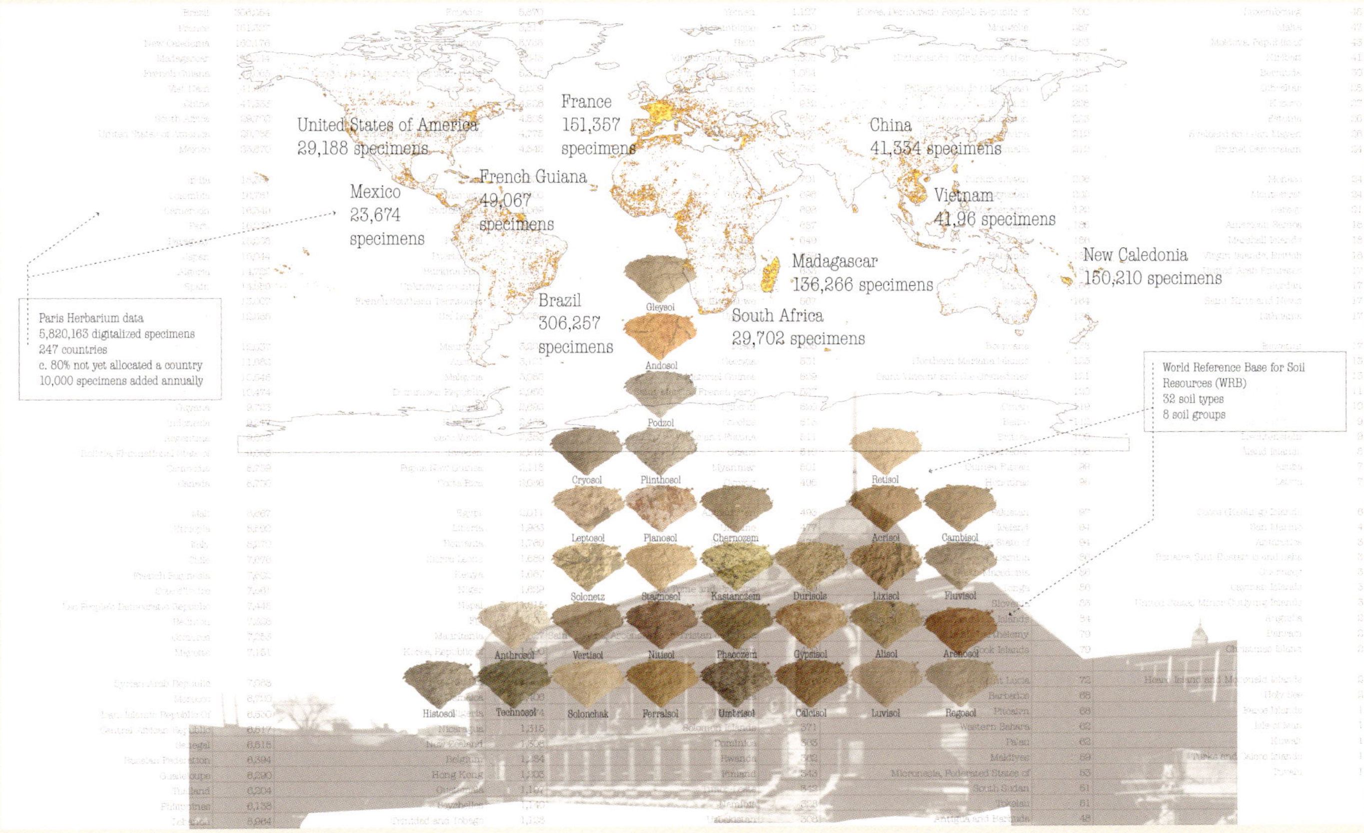
United States of America
29,188 specimens
France
151,357
specimens
China
41,334 specimens
Mexico
23,674
specimens
French Guiana
49,067
specimens
Vietnam
41,96 specimens
New Caledonia
150,210 specimens
Madagascar
136,266 specimens
Brazil
306,257
specimens
South Africa
29,702 specimens
Paris Herbarium data
5,820,163 digitalized specimens
247 countries
c. 80% not yet allocated a country
10,000 specimens added annually
World Reference Base for Soil Resources (WRB)
32 soil types
8 soil groups
Gleysol
Andosol
Podzol
Cryosol
Plinthosol
Retisol
Leptosol
Planosol
Chernozem
Acrisol
Cambisol
Solonetz
Stagnosol
Kastanozem
Durisols
Lixisol
Fluvisol
Anthrosol
Vertisol
Nitisol
Phaeozem
Gypsisol
Alisol
Arenosol
Histosol
Technosol
Solonchak
Ferralsol
Umbrisol
Calcisol
Luvisol
Regosol

ALLÉE OF DARKNESS

LOUISE BANI SARCAR

We explore what we want to see; what we explore is what we conquer, and once we conquer, we exploit. This repetitive cycle of visual to material consumption has resulted in the world we live in today.

During the Renaissance, the nobility held power over their territories from their palaces and villas, which provided a commanding view of the surrounding landscape. Much of the dominance of the human species depends on this power of sight, and those in power choreograph what they want others to see. This was especially true of the grand, axial landscapes of baroque France. In our proposal, the long-standing gaze of power along the Jardin des Plantes is blocked with towering walls. An allée of darkness is created to express the sublimity of natural elements as opposed to the tamed and controlled versions of nature in the adjacent gardens. Descending into the forecourt from the greenhouse, one must navigate this dark allée by relying on other senses. The sound of a waterfall acts as a wayfinding guide, directing people from one point to the next. Where the walls narrow, one can touch the rocky patina. At other points along the path, the walls widen, and one finds a sliver of opening to enter the forecourt. Perhaps by being enveloped in darkness and attuning to the world through non-visual senses, we can destabilize what we think we know, opening new relationships based on empathy and connection.

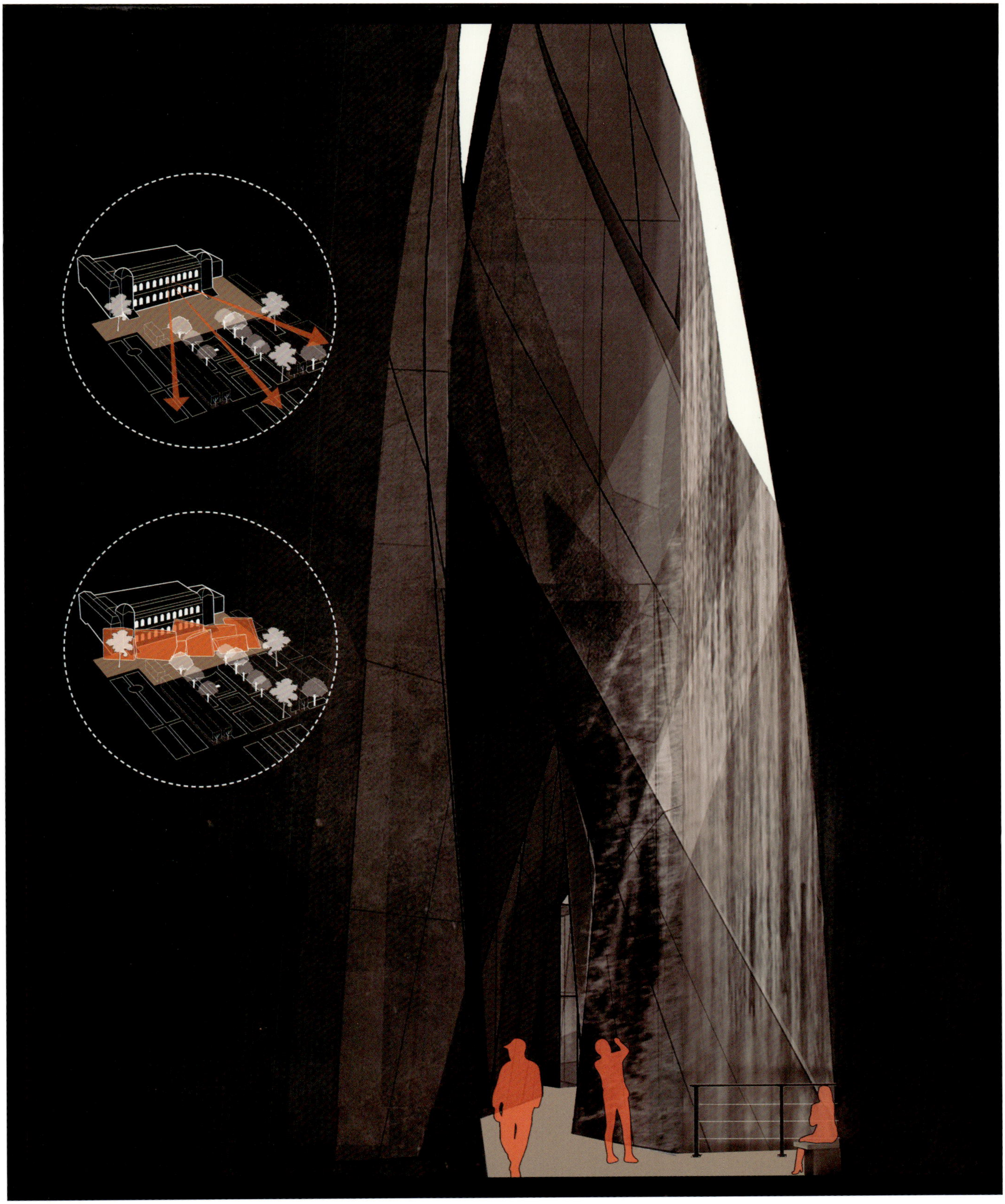

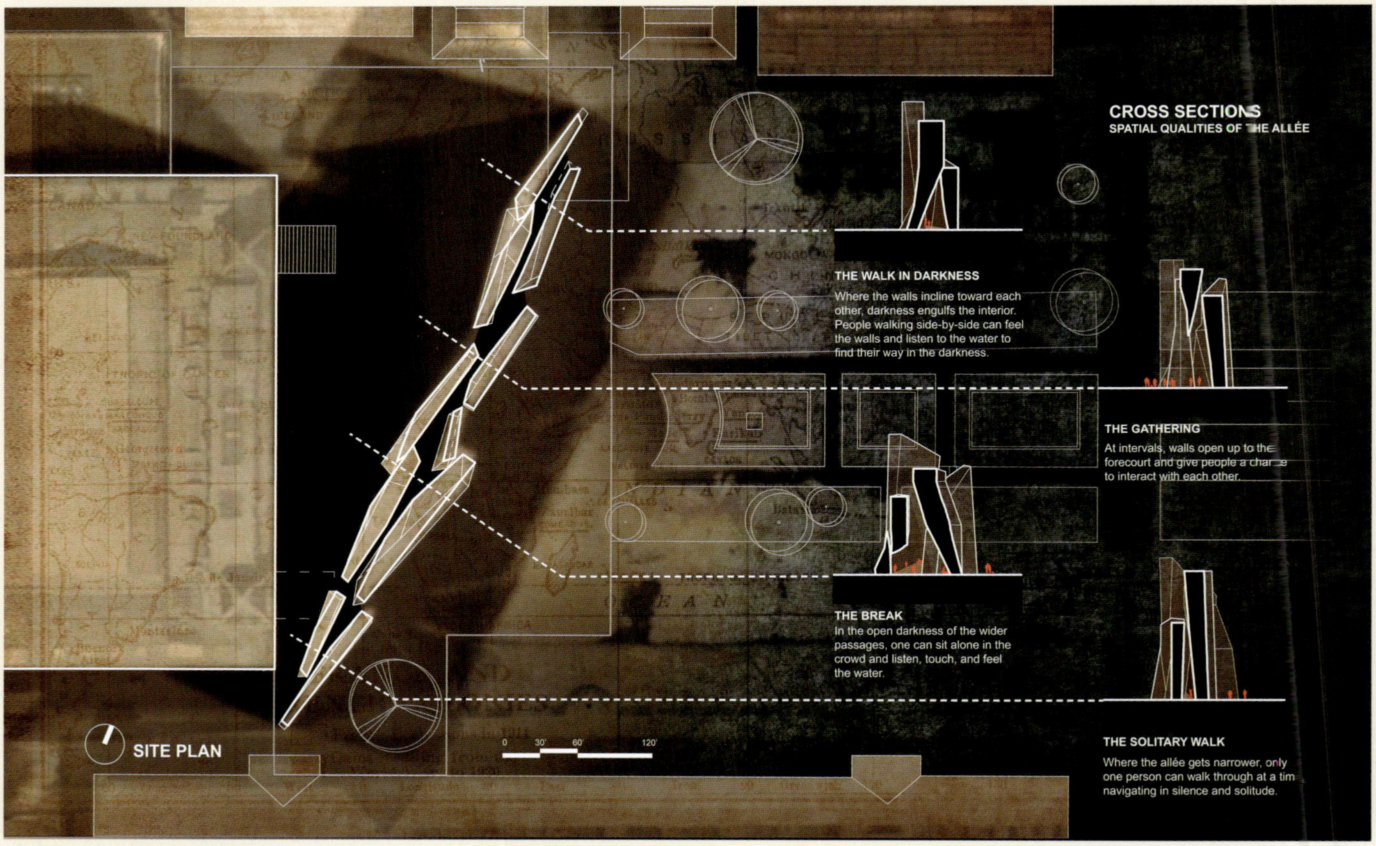
CROSS SECTIONS
SPATIAL QUALITIES OF THE ALLÉE
THE WALK IN DARKNESS
Where the walls incline toward each other, darkness engulfs the interior. People walking side-by-side can feel the walls and listen to the water to find their way in the darkness.
THE GATHERING
At intervals, walls open up to the forecourt and give people a chance to interact with each other.
THE BREAK
In the open darkness of the wider passages, one can sit alone in the crowd and listen, touch, and feel the water.
THE SOLITARY WALK
Where the allée gets narrower, only one person can walk through at a time navigating in silence and solitude.
SITE PLAN
0 30' 60' 120'

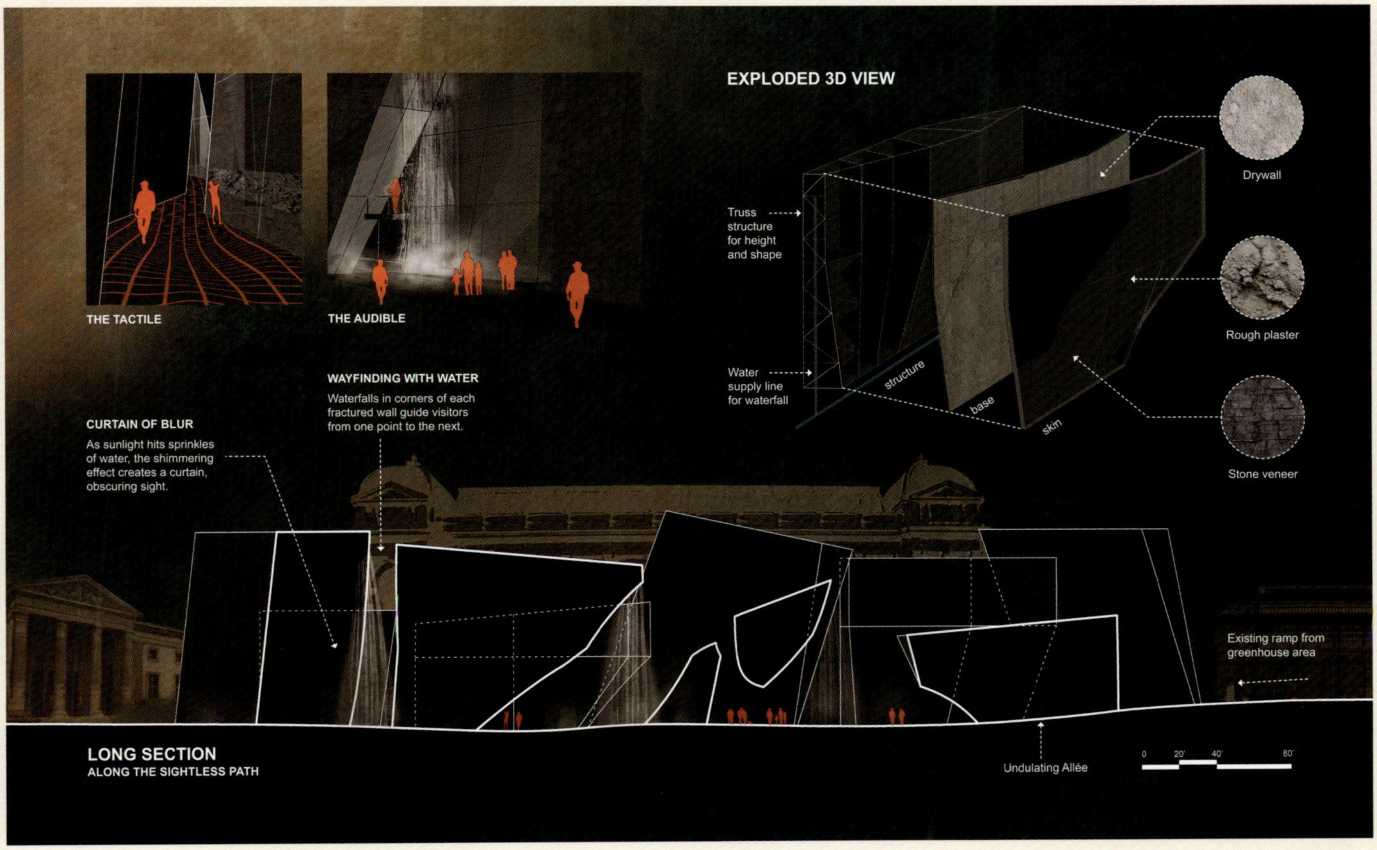
EXPLODED 3D VIEW
Drywall
Truss structure for height and shape
Rough plaster
Water supply line for waterfall
structure
base
skin
Stone veneer
THE TACTILE
THE AUDIBLE
WAYFINDING WITH WATER
Waterfalls in corners of each fractured wall guide visitors from one point to the next.
CURTAIN OF BLUR
As sunlight hits sprinkles of water, the shimmering effect creates a curtain, obscuring sight.
Existing ramp from greenhouse area
LONG SECTION
ALONG THE SIGHTLESS PATH
Undulating Allée
0 20' 40' 80'

INTANGIBLE ARCHIPELAGO

WAI LO CIARA

In today's world of globalized trade, colonized cultures, and homogenized ecosystems, the concept of the exotic extends beyond physical objects and artifacts. It also includes abstract notions, ideologies, and perspectives of people considered "others" that still captivate and intrigue us. In contrast to the existing exhibition halls at the Jardin des Plantes, which resemble time-capsule Wardian cases containing tangible discoveries from the past, this proposal encapsulates both the essence of the tangible past (through the selection of plant species) and the intangible present of the "exotic" as it evolves and is cultivated through new intellectual and ecological connections.

Intangible Archipelago comprises seven "islands," representing seven continents, planted with once-exotic species that were widely cultivated and have now become an integral part of France's current landscape, even serving important sociocultural roles. These acclimated exotic plants coexist symbiotically with native species and pollinators in Paris. The composition of islands converges toward the center, forming a circular platform for gatherings: a place to celebrate and appreciate others' ideologies. The coming-together of the islands goes beyond the symbolic: it also nurtures connections within the underground fungal network of each island with its neighboring biome.

Intangible Archipelago

The archipelago comprises 7 islands, each symbolizing a continent, forming a circular intellectual platform. It is home to 7 acclimated exotic French plants living symbiotically with native species and pollinators in Paris.

Beyond celebrating the appreciation of others' ideologies, it also nurtures connections within the underground fungal network of the islands and with the neighboring biome.

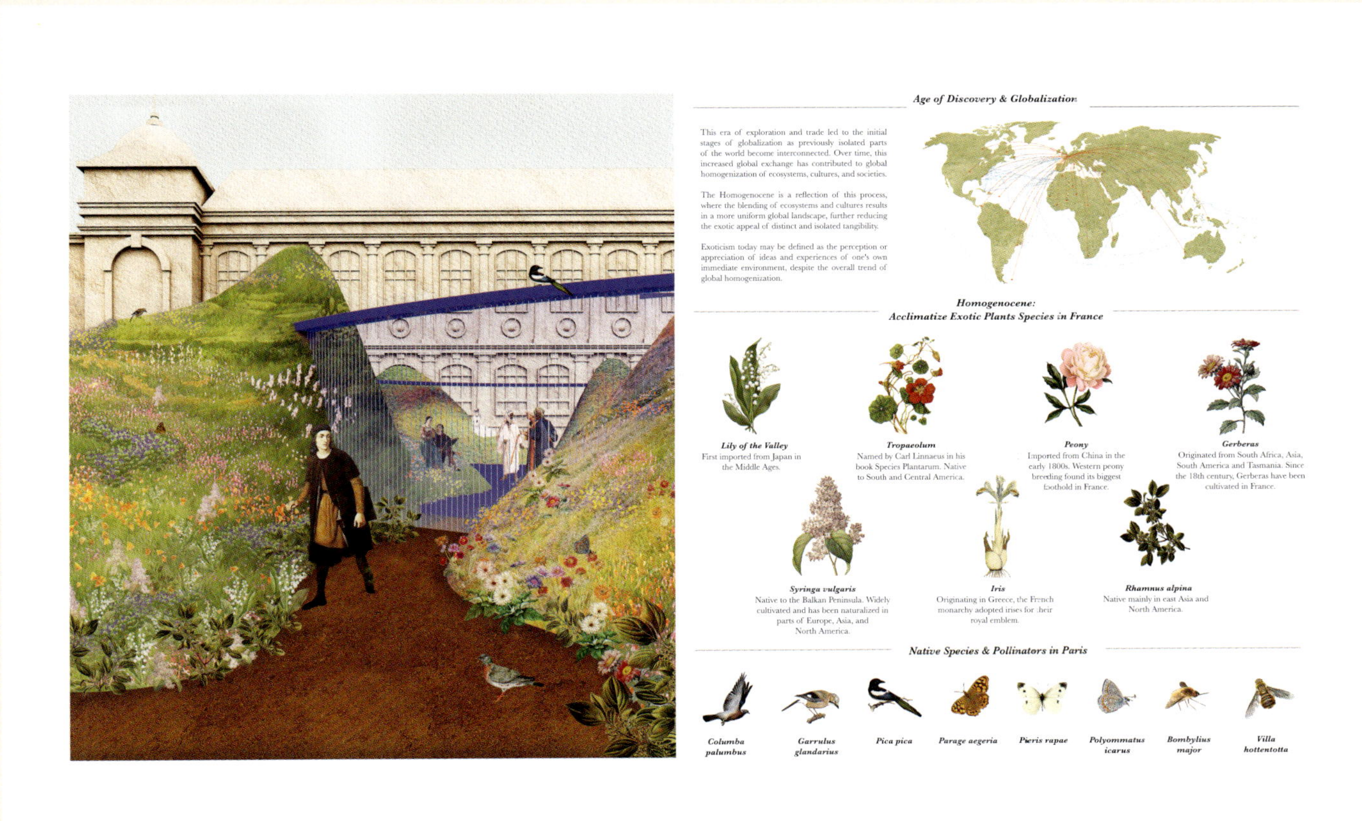
Age of Discovery & Globalization
This era of exploration and trade led to the initial stages of globalization as previously isolated parts of the world become interconnected. Over time, this increased global exchange has contributed to global homogenization of ecosystems, cultures, and societies.
The Homogenocene is a reflection of this process, where the blending of ecosystems and cultures results in a more uniform global landscape, further reducing the exotic appeal of distinct and isolated tangibility.
Exoticism today may be defined as the perception or appreciation of ideas and experiences of one's own immediate environment, despite the overall trend of global homogenization.
Homogenocene:
Acclimatize Exotic Plants Species in France
Lily of the Valley
First imported from Japan in the Middle Ages.
Tropaeolum
Named by Carl Linnaeus in his book Species Plantarum. Native to South and Central America.
Peony
Imported from China in the early 1800s. Western peony breeding found its biggest foothold in France.
Gerberas
Originated from South Africa, Asia, South America and Tasmania. Since the 18th century, Gerberas have been cultivated in France.
Syringa vulgaris
Native to the Balkan Peninsula. Widely cultivated and has been naturalized in parts of Europe, Asia, and North America.
Iris
Originating in Greece, the French monarchy adopted irises for their royal emblem.
Rhamnus alpina
Native mainly in east Asia and North America.
Native Species & Pollinators in Paris
Columba palumbus
Garrulus glandarius
Pica pica
Parage aegeria
Pieris rapae
Polyommatus icarus
Bombylius major
Villa hottentotta

DE | COLONIZE

CLAIRE NAPAWAN • LINDA CHAMORRO • MARC MILLER

To plant a garden is to enter a relationship with the natural world. Many botanic gardens represent a process of colonization, extraction, and exotification. The plants displayed at the Jardin des Plantes are consistent with the narrative that supports the "othering" of plants, places, and cultures outside the West. However, the Jardin des Plantes is also the site of a seed bank actively working to maintain seed mixes that are native and adapted to the local microclimatic conditions of Paris.

Our proposal seeks to de|colonize past notions of what a plant collection can be by employing a participatory and opportunistic planting system. Thus, the exotic becomes not the specimen but, rather, how the garden form is determined. Troughs of planting beds are dug into the decomposed granite surface of the museum's forecourt, flush with the existing grade, and mulched with matching decomposed granite. Flush trays filled with moistened native seed mixes are placed at each entry to the plaza. As visitors enter, the seeds adhere to the soles of their shoes and are dispersed throughout the decomposed granite forecourt of the museum. Seeds take root in the locations that support their colonization of the site. As the opportunistic plants grow, visitors help monitor what germinates and thrives – or what does not. The native and adaptive plants in the forecourt represent a dramatic foil to the surrounding manicured and maintained gardens and the precious treatment of specimens within the museum. This represents a new relationship with the natural world that allows for constant evolution and embraces the "messy."

MUSÉUM
NAL HIST
NATURELLE

1
2
3
4
MUSEUM NATIONAL D'HISTOIRE NATURELLE
THE DE | COLONIZE PROCESS:
1. TROUGHS OF PLANTING BEDS ARE DUG INTO THE DECOMPOSED GRANITE SURFACE OF THE MUSEUM'S FORECOURT, FLUSH WITH EXISTING GRADE AND MULCHED WITH MATCHING DECOMPOSED GRANITE.
2. FLUSH TRAYS FILLED WITH MOISTENED NATIVE SEED MIXES ARE PLACED AT EACH ENTRY TO THE PLAZA.
3. AS VISITORS ENTER, THE SEEDS ADNERE TO THE SOLES OF THEIR SHOES AND ARE DISPERSED THROUGHOUT THE DECOMPOSED GRANITE FORECOURT OF THE MUSEUM.
4. SEEDS TAKE ROOT IN THE LOCATIONS THAT SUPPORT THEIR COLONIZATION OF THE SITE.
5. AS THE OPPORTUNISTIC PLANTS GROW, VISITORS HELP MONITOR WHAT GERMINATES, THRIVES, OR FAILS TO TAKE ROOT.

museum_goer_2024
FOLLOW
357 likes
museum_goer_2024 check out these crazy plants! #exotique
nativeplantlover why are there WEEDS at the jardin de plantes????

jadoreplantes
FOLLOW
765 likes
jadoreplantes plantes opportunistes tres chic
grouprojects #exotique #laplus
citizen_scientist interesting to see #nativeplants at botanical collection!

novel_ecologies
FOLLOW
7k likes
novel_ecologyies native as #exotique how novel
ucdavis_iaed #decolonizingplants #decoloniznglandscapes

EXPLORATION GROUNDS

DALE WIEBE • RYAN COATES

The collections of living matter archived in the Muséum national d'Histoire naturelle are representations of distant landscapes. Our proposal pays homage to the grounds left behind, translating the geological grounds of research sites from the past into a unified sloping surface representing these sites of extraction. The terrain of the forecourt is lifted to meet the existing entrance level of the museum and lowered where it meets the garden. Modifying the topography allows universal access to the museum and limited access to the existing garden. The inaccessible garden becomes an island, left to evolve on its own.

Abstract representations of the grounds of foreign research sites are composed of limestone at varying stages of erosion, from rocky outcrop to aggregate. This material references the local area's history of limestone quarries that produced the distinguished architectural stone, "Paris Stone." The limestone surface has various textures that encourage the deposition of organic matter and the spontaneous seeding of a new landscape. Through the "collection" of distant grounds, the project aims to highlight the tension between how we organize living matter and the maintenance practices of a living archive.

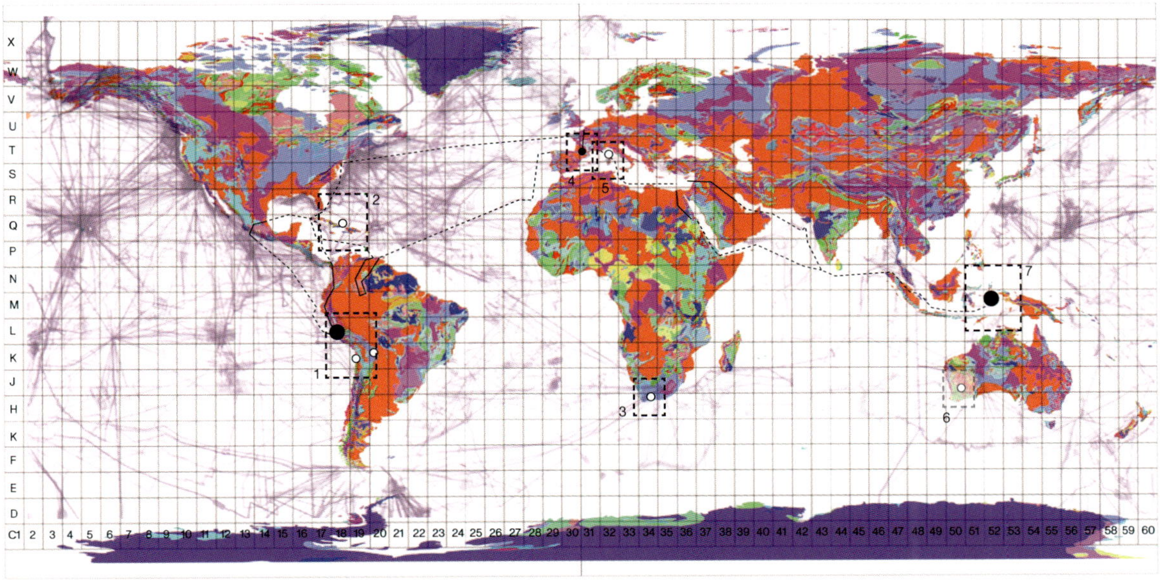

Early Expeditions

(line)	Pierre Sonnerat	Voyage à la Nouvelle-Guinée (1776)
●	Alexander Von Humboldt	Humboldt's Voyage aux régions équinoxiales du Nouveau Continent (1820)
	Édouard Lartet	New Research on the Coexistence of Man and of the Great Fossil Mammifers characteristic of the Last Geological Period (1861)

Active Research

○	CAFOTROP program	Tropical Rainforest, study of evolution of species from the 'supercontinent' called Gondwana.
	Tropical Deep-Sea Benthos	Deep sea collection and research of marine organisms.
	La Planète Revisitée program	Inventory of Algae, marine invertebrates and terrestrial on Corsica. (2019-2022)
(line)	Multibeam Bathymetric Surveys	Diverse expeditions surveying ocean floor

General geological

A - Cenozoic sedimentary	Von Humboldt, Lartet expedition.
B - Paleozoic sedimentary	Alexander Von Humboldt expedition, CAFOTROP program
C - Mesozoic sedimentary	Alexander Von Humboldt expedition, CAFOTROP program
D - Mesozoic-Cenozoic volcanic	Alexander Von Humboldt expedition.
E - Cenozoic volcanic rocks	Sonnerat Expedition
F - Paleozoic intrusive	La Planète Revisitée program
G - Archean crystalline metamorphic	CAFOTROP program

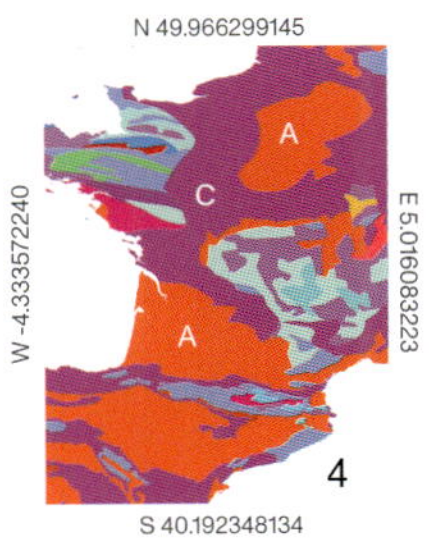

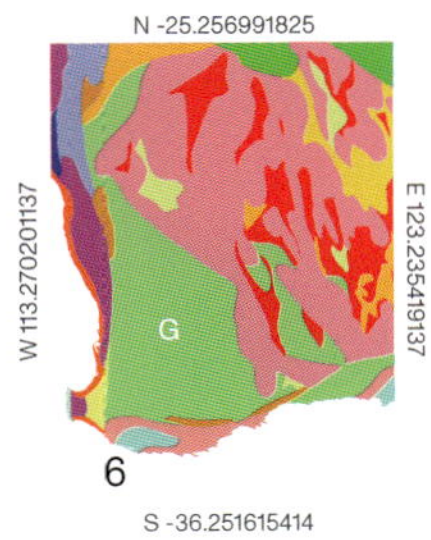

Reseach Site

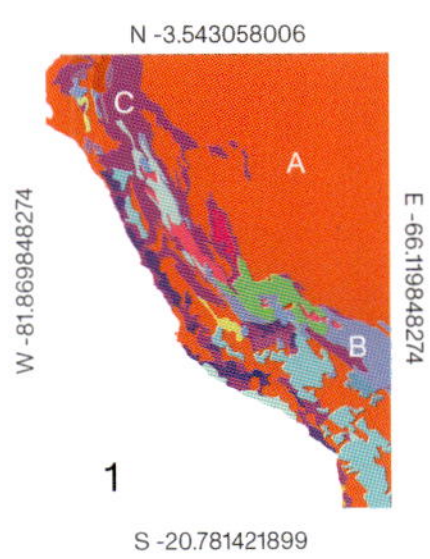

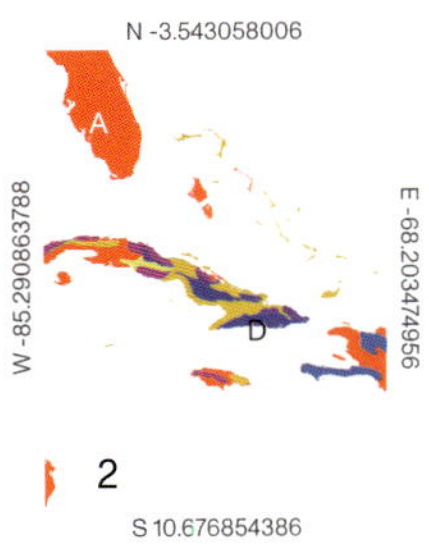

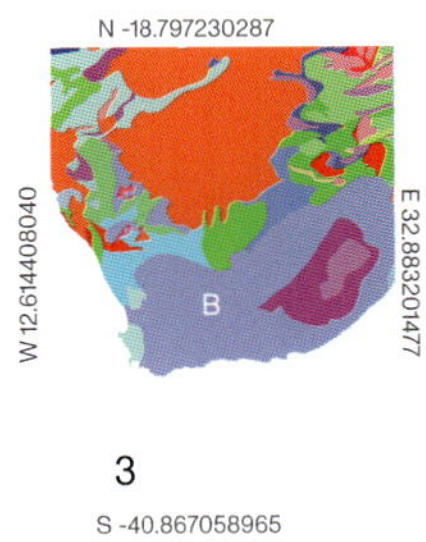

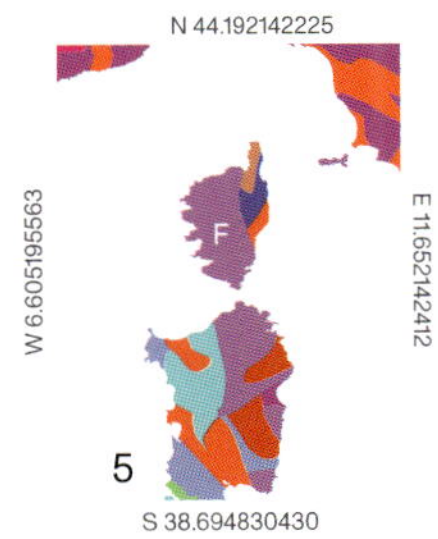

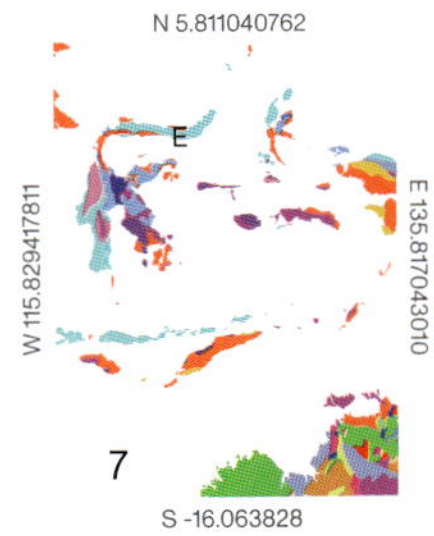

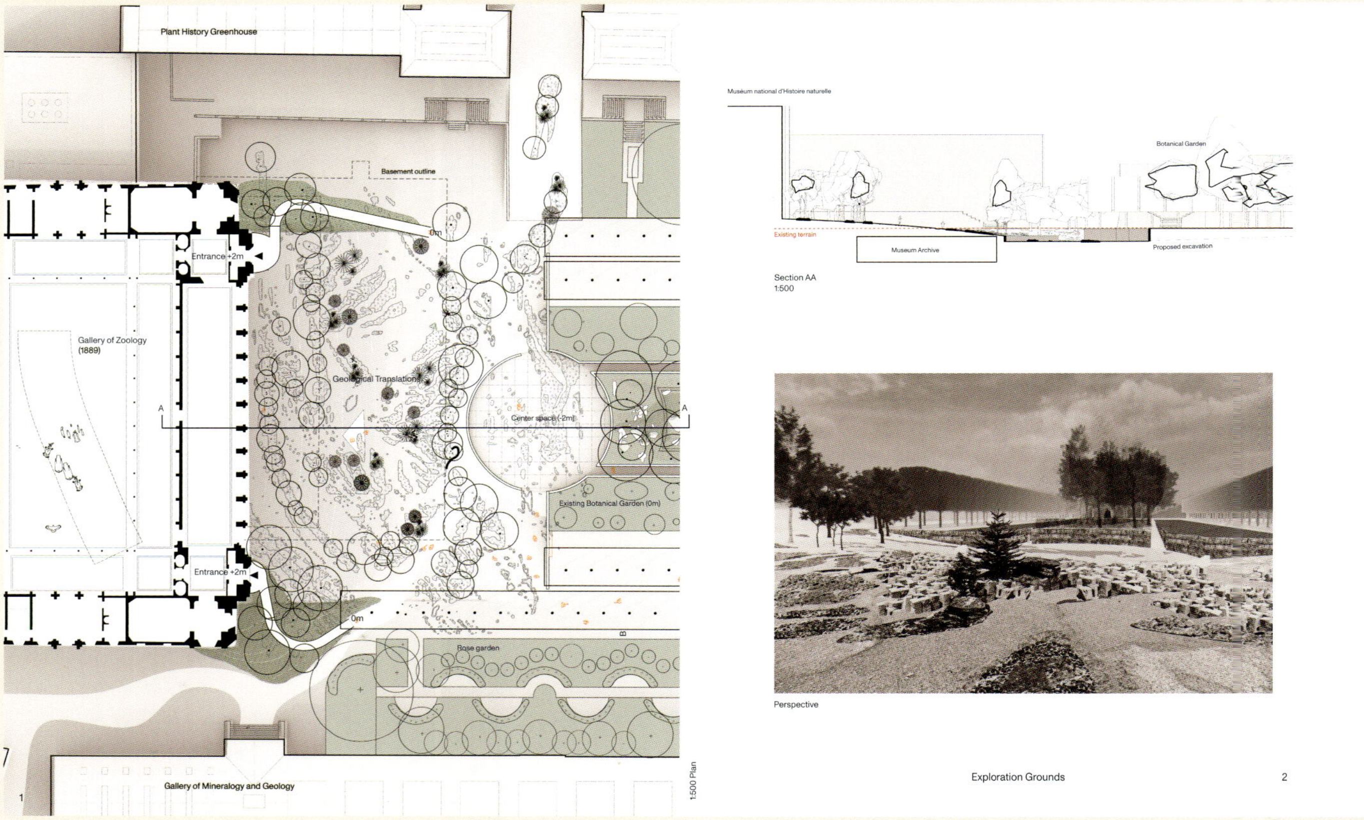
Plant History Greenhouse
Basement outline
0m
Entrance +2m
Gallery of Zoology
(1889)
Geological Translations
A
Center space (-2m)
A
Existing Botanical Garden (0m)
Entrance +2m
0m
Rose garden
Gallery of Mineralogy and Geology
1
1:500 Plan
Muséum national d'Histoire naturelle
Botanical Garden
Existing terrain
Museum Archive
Proposed excavation
Section AA
1:500
Perspective
Exploration Grounds
2

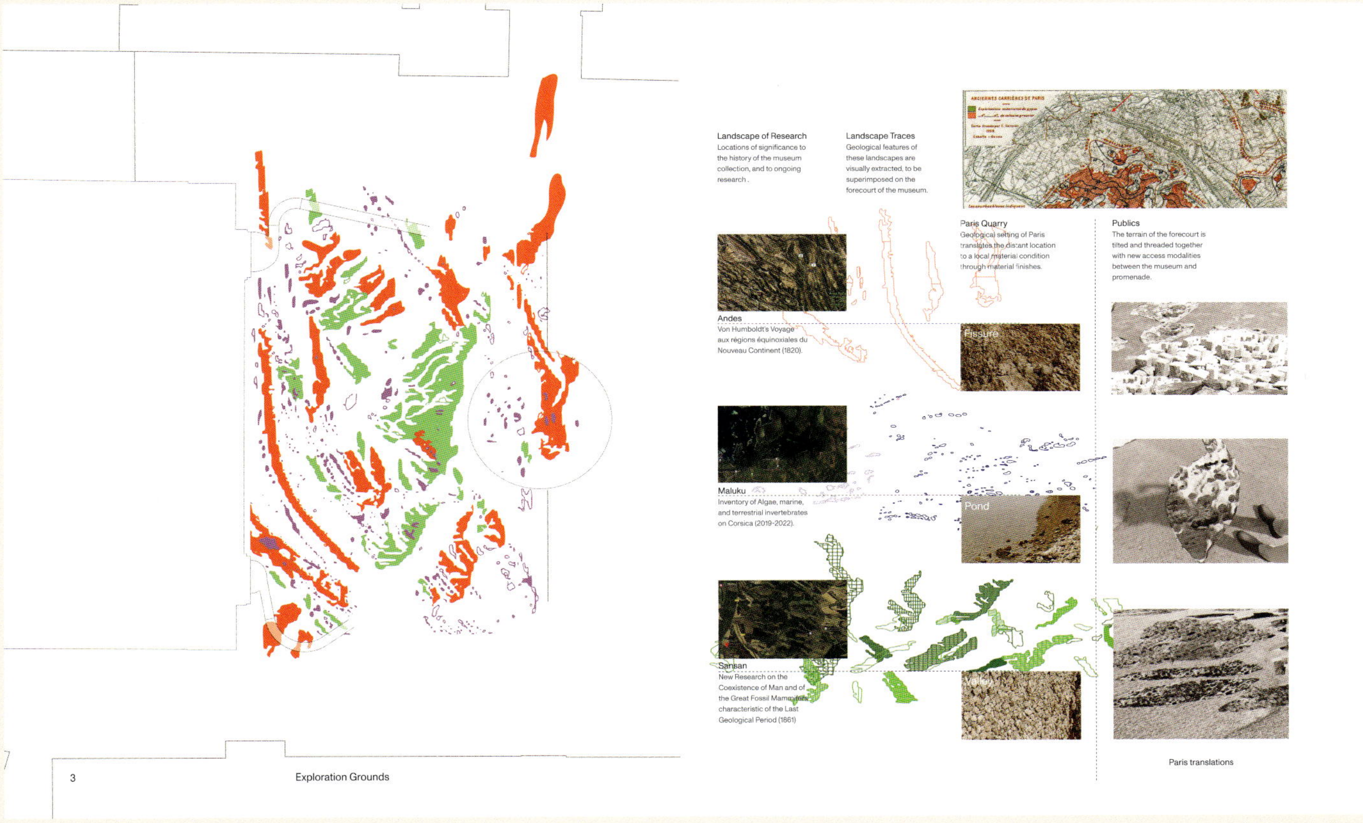
3
Exploration Grounds
Landscape of Research
Locations of significance to the history of the museum collection, and to ongoing research .
Landscape Traces
Geological features of these landscapes are visually extracted, to be superimposed on the forecourt of the museum.
ANCIENNES CARRIÈRES DE PARIS
Paris Quarry
Geological setting of Paris translates the distant location to a local material condition through material finishes.
Publics
The terrain of the forecourt is tilted and threaded together with new access modalities between the museum and promenade.
Andes
Von Humboldt's Voyage aux régions équinoxiales du Nouveau Continent (1820).
Fissure
Maluku
Inventory of Algae, marine, and terrestrial invertebrates on Corsica (2019-2022).
Pond
Sansan
New Research on the Coexistence of Man and of the Great Fossil Mammifers characteristic of the Last Geological Period (1861)
Valley
Paris translations

SALON DES REFUSÉS

20
30
40
50
60
70
80
90
100
110
120
130
Gallery of Evolution
Gallery of Mineralogy
Gallery of Botany
Gallery of Paleontology
Greenhouses
Garden of Possibilities
Museum Archive
Terrarium Gallery
Cave of Explorations
Gallery of Evolution
Marl
Paris's Quarry
Sand

Le Grand Bazar de l'Existence by Arthur Perez + Nicole Wagy: The project challenges the process, preservation, and organization of novel findings from far-off places. A sequence of revelatory experiences prepares visitors to learn through contrasting themes: the known and the unknown, the past and the future, the controlled and the wild.

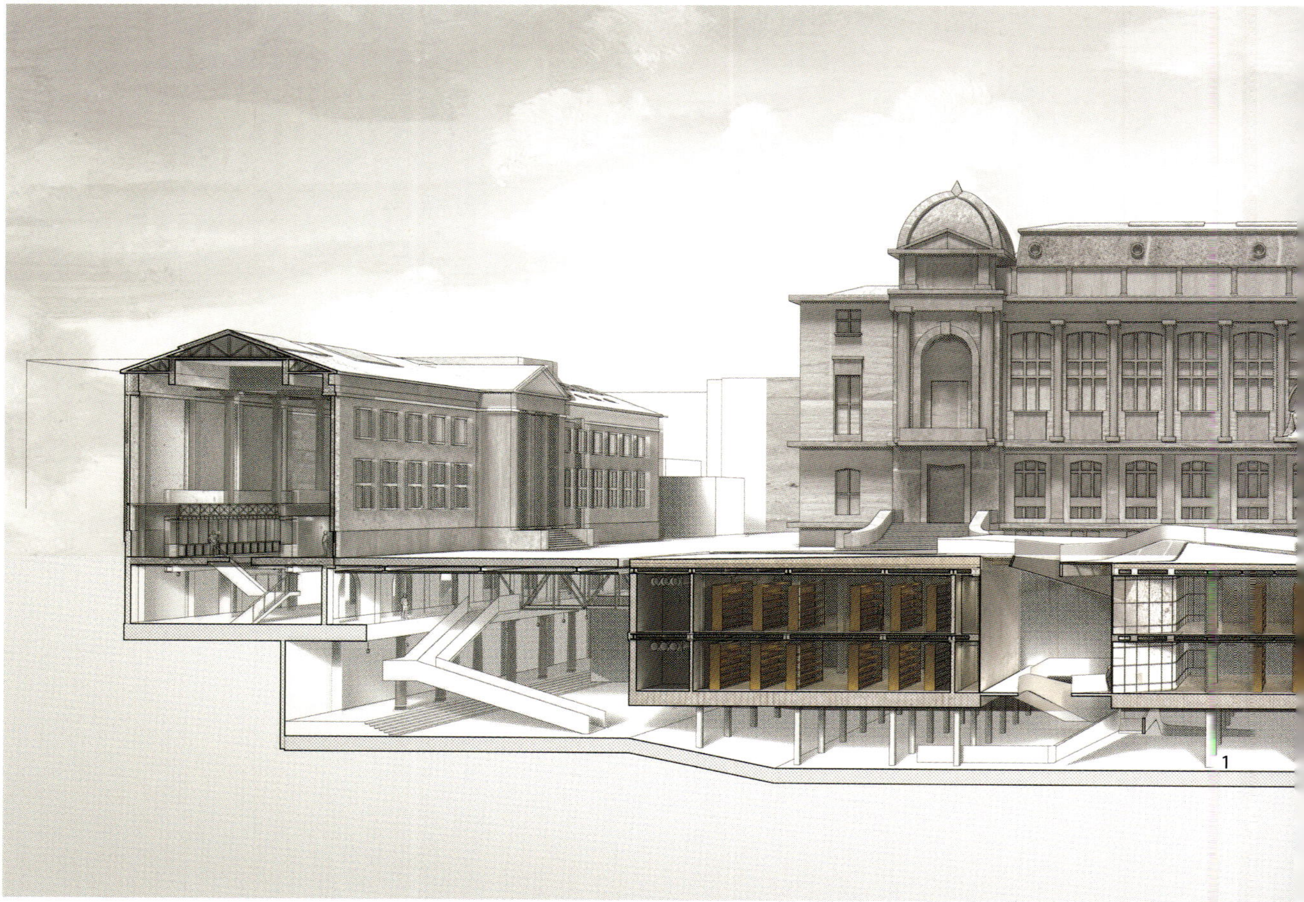
1

Arkheion – Memorial of Archive by Runkai Jin: Who decides what information should be preserved and what should be forgotten? A surface "crack" in the museum's forecourt reveals the existing archive beneath it, drawing visitors' attention and forming new access and a new program: the exhibition of the archive.

Statue of Lamark
[further down 250M]
Vertebrata
Statue of Buffon
Abies alba
European silver fir
Quercus pubescens
downy oak
Carpinus betulus
European hornbeam
Fraxinus excelsior
European ash
Taxus baccata
the Yew
Grande Galerie de l'Évolution
4500mm
9000mm
Cross Section
Existing Basement
Sealed Basement

1

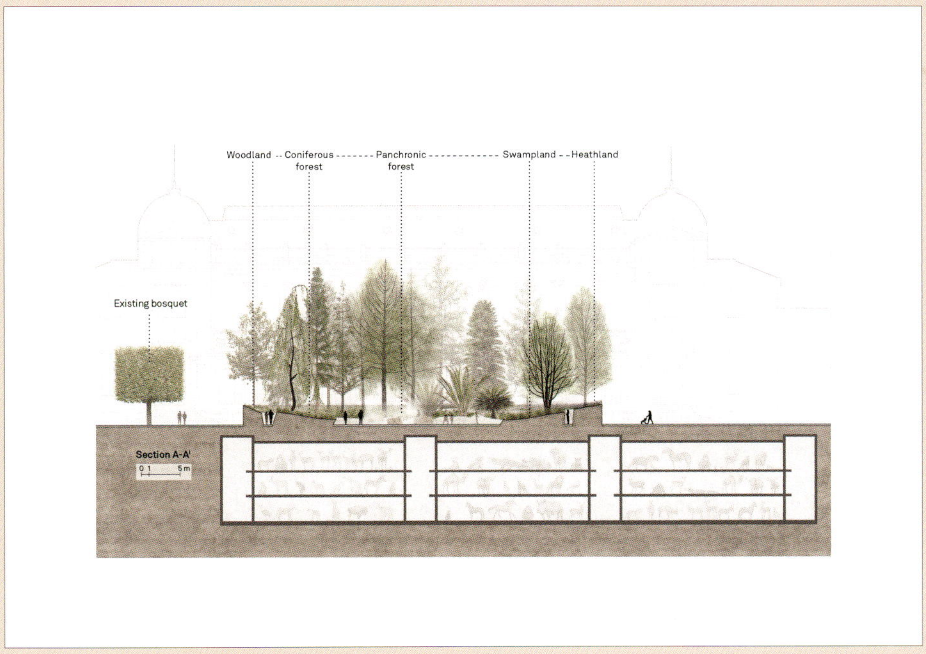

2

1. The Garden Axis Evolution Project by Jiacheng Chen + Xiaohang Wu: Sealed with native tree species from France, nine exotic stupa structures adapted from architectural icons of Laos and Myanmar (two former French colonies) are aligned with the heroic statues of Buffon and Lamarck, enriching the symbolic meaning of the axis in the Jardin des Plantes.

2. Hortus Lutetiensis by Magnus Hehlke: The garden addresses the planet's ongoing biological and geologic processes by first highlighting the Lutetian landscape and geological forms of the past, then providing insight into the future Parisian landscape.

IV
CHAPTER
While with an eye made quiet by the power Of harmony,
and the deep power of joy,
we see into the life of things.
William Wordsworth

III
CHAPTER
In the darkness, nature reveals her mysterious side,
allowing us to more humbly appreciate her greatness.
David Saunders

Déjà vu Garden by Qianhe Xu, Yiwei Chen + Chen Bo: The Déjà vu Garden is an underground garden of dreamlike unconventional natural spaces. In the subterranean expanse, we aim to awaken dormant perceptions of nature.

INTRODUCING THE KEY PLAYERS

ALLYSHIP

The Allyship is a public entity made up of non-Assembly Members who are committed to supporting the Stateless Assembly and its efforts

Allyship

Allyship Sentinels

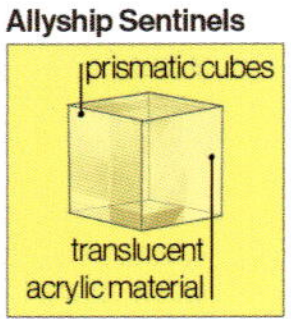

Structures

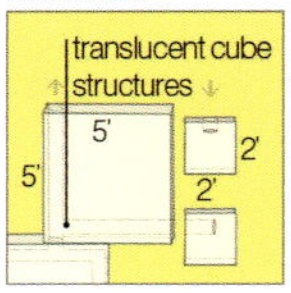

Lighting

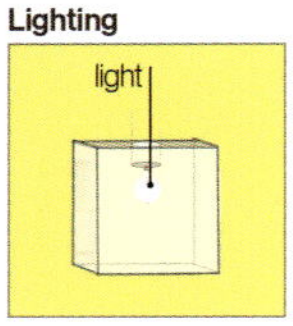

STATELESS ASSEMBLY AND SITES

When in session, Assembly voices are transcribed as water ripples in the Sites' pools, creating a public *presence* for the Assembly while allowing its Members to remain anonymous and decentralized

Decentralized Assembly

Onsite 'Voice' Ripples

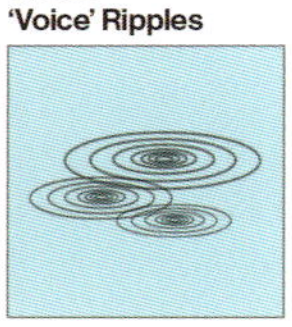

DISRUPTING COLONIAL POWER

The Assembly Site at The Gallery of Evolution's forecourt unearths the museum's existing underground storage facility, revealing hidden colonial power structures to the public

Museum Storage Underneath Forecourt

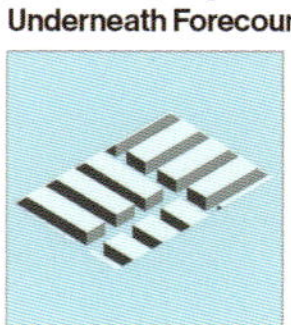

STATELESS ASSEMBLY SITE

AT THE GALLERY OF EVOLUTION FORECOURT

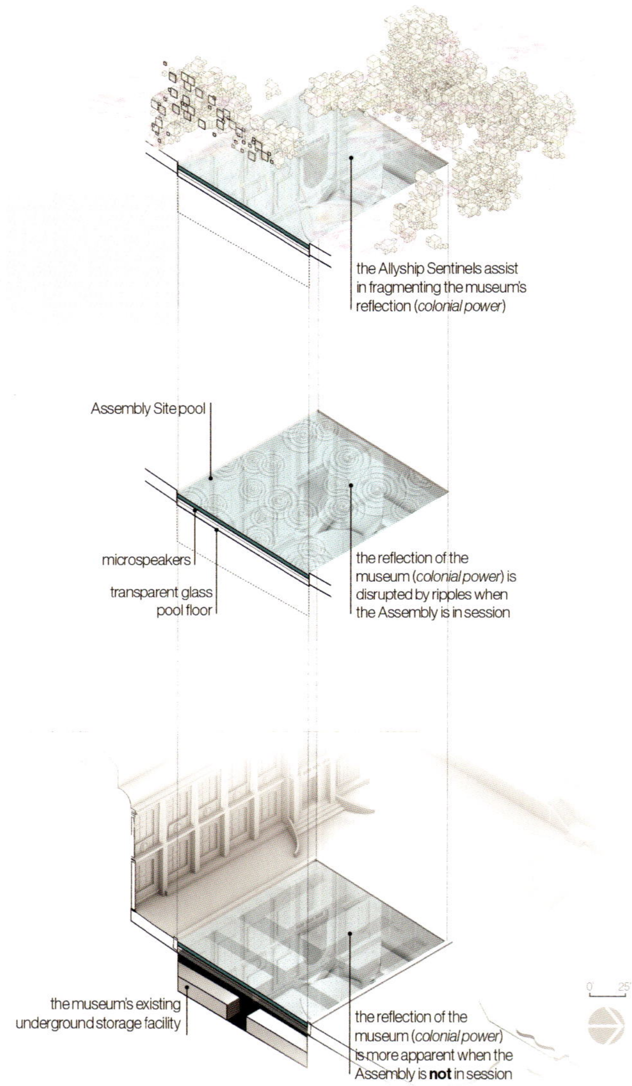

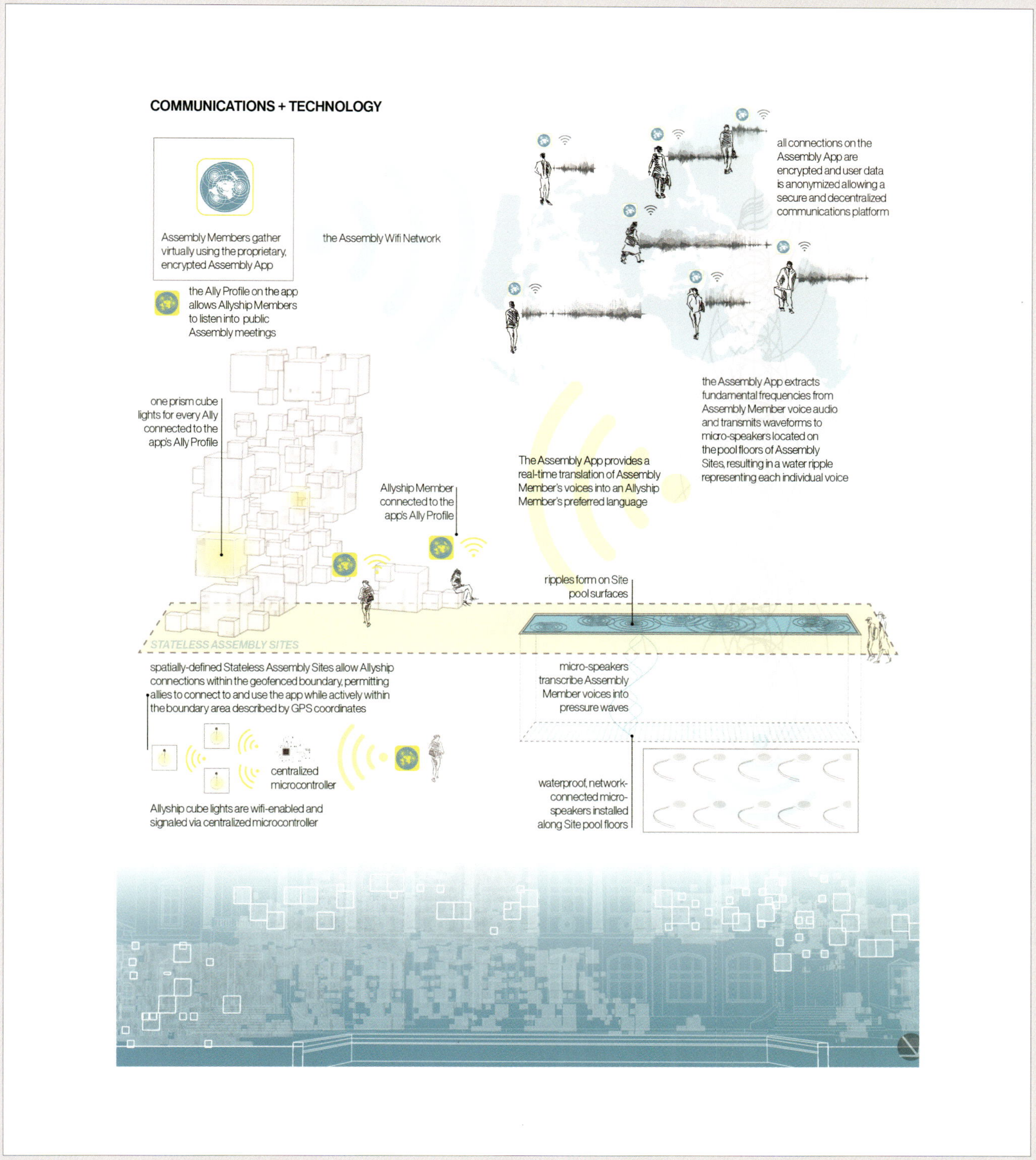

The Stateless Assembly by Thea Swift: The Stateless Assembly is an international governing body seeking justice for victims caught in the reverberating and relentless states of unbelonging and colonial violence.

1

2

1. The Digital Garden by Sungsu Kang:
Exoticism began by portraying the East as a mysterious and conquerable world. The real plantings in the "Garden of Present" contrast with the "Garden of Extinction," composed of LED screens portraying lost plants of the East.

2. Court of Reflection by Yujue Wang:
The "Court of Reflection" sculptures remain in the historical fence's footprint, having been reappropriated with up-to-date museum posters. These sculptures are open for visitors to freely express themselves, fostering a space for reflection and dialogue.

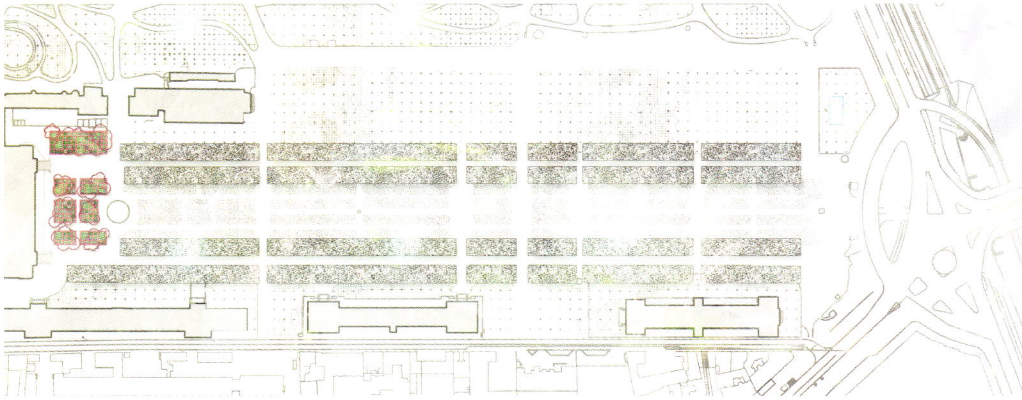

1

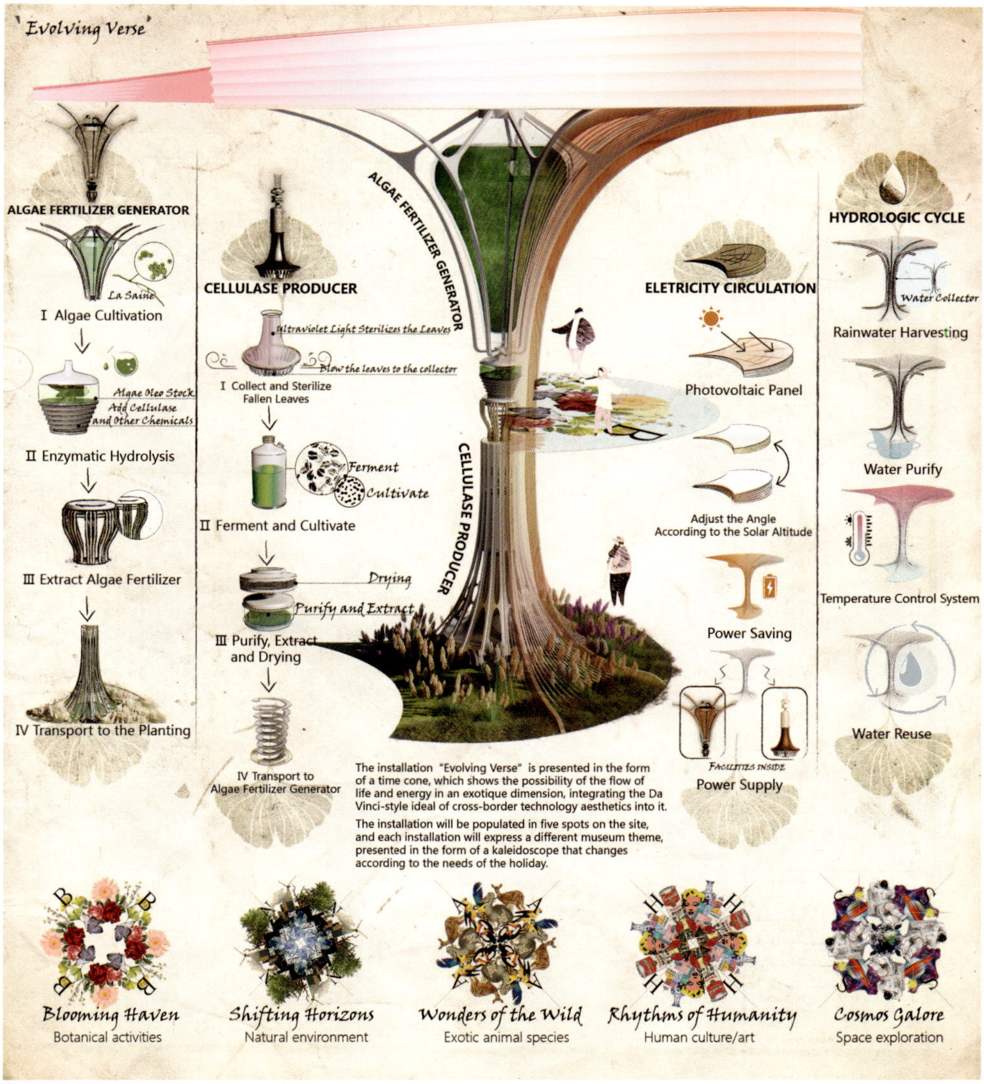

2

1. The Migration Microforest by Ilana Cohen:
This proposal transforms the museum forecourt into a densely planted experimental forest plot that tests the adaptability of various exotic plant species to the changing climate of Paris.

2. Enigmatic Dimension by Zikun Zhang, Songyi Zheng + Lingyi Xu:
Considering geologic time scales, the emergence of human beings and their influence on the environment is a sudden, "exotic" condition for nonhuman species. The ancient ginkgo tree is used as a symbolic element connecting the past, present, and future.

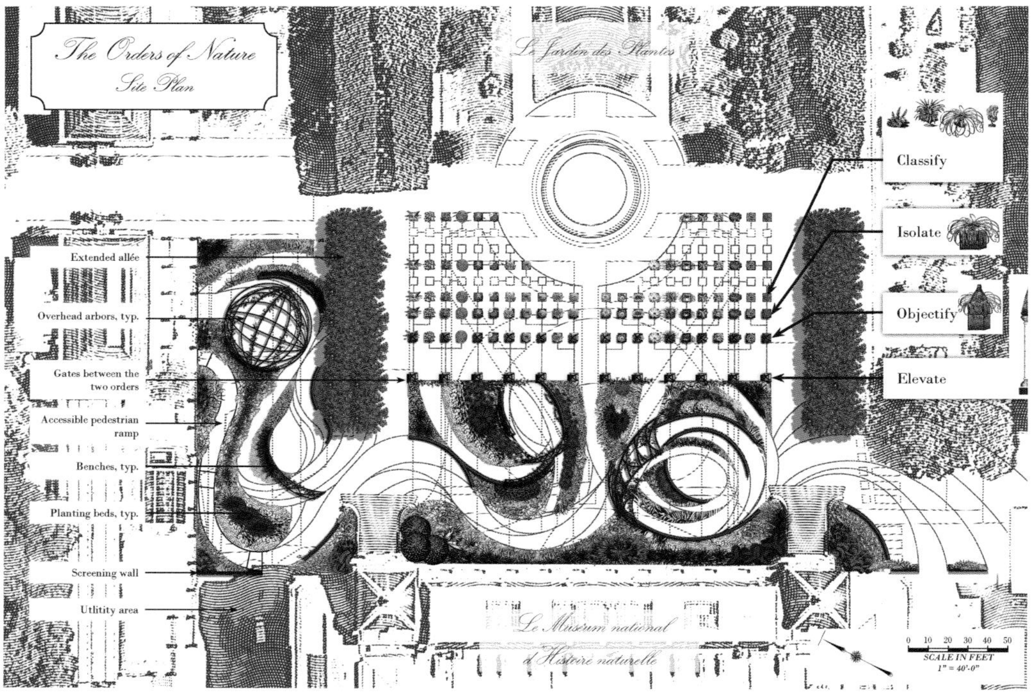
The Orders of Nature
Site Plan
Le Jardin des Plantes
Classify
Isolate
Objectify
Elevate
Extended allée
Overhead arbors, typ.
Gates between the two orders
Accessible pedestrian ramp
Benches, typ.
Planting beds, typ.
Screening wall
Utlitity area
Le Muséum national
d'Histoire naturelle
0 10 20 30 40 50
SCALE IN FEET
1" = 40'-0"

The Orders of Nature by Katherine Melcher: Taxonomic order is juxtaposed with "vital entanglement," inspired by the dynamic and unfolding connections among all living beings. Warped grids, unfurling vines, and multiple, overlapped ordering systems beg the question: Who is ordering whom?

1.35 m
2.7 m

EDITOR'S CHOICE

The Parterre of Topiary Anamorphosis by Yufan Gao: The proposal features distorted botanical patterns viewed through green conical mirrors resembling topiaries, using anamorphism to symbolize the relativity of vision and to challenge the narrative in museum displays that often place one perspective at the center.

IN THE NEXT ISSUE OF

LA+

Ever more technologies are being created to sense our environment, and much is being learned about how animals and plants sense theirs. We often think of these tools as extending our capacity for sensing what is not available through natural human perception. But what is "natural" about human perception? Not as much as was once believed, it turns out. Though our senses have a biological basis, they are not simply intermediaries through which we gain empirical knowledge about our world. How do senses become naturalized, and bodies and experiences standardized? Find out in our next issue, **LA+ SENSE,** with contributors reflecting on how we understand our world through various senses, sensors, and sensibilities.

ELENA GIULIA ABBIATICI
SARAH COLEMAN
TIM CRESSWELL
LISA YIN HAN
AI HISANO
DAVID HOWES
MARK KINGWELL
JIA HUI LEE
GASCIA OUZOUNIAN
KRIS PAULSEN
SALLY PUSEDE
ERIN PUTALIK
DOUGLAS ROBB
CHRIS SALTER
ALEXA VAUGHN
ALEXA WEIK VON MOSSNER
MARK PETER WRIGHT

OUT **SPRING 2025**

WILD SPRING 2015 | PLEASURE FALL 2015 | TYRANNY SPRING 2016

SIMULATION FALL 2016 | IDENTITY SPRING 2017 | RISK FALL 2017

IMAGINATION SPRING 2018 | TIME FALL 2018 | DESIGN SPRING 2019

ICONOCLAST FALL 2019 | VITALITY SPRING 2020 | GEO FALL 2020

COMMUNITY SPRING 2021 | CREATURE FALL 2021 | GREEN SPRING 2022

LA+
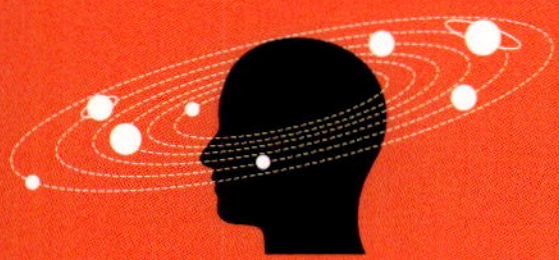
SPECULATION FALL 2022

LA+

INTERRUPTION SPRING 2023

LA+
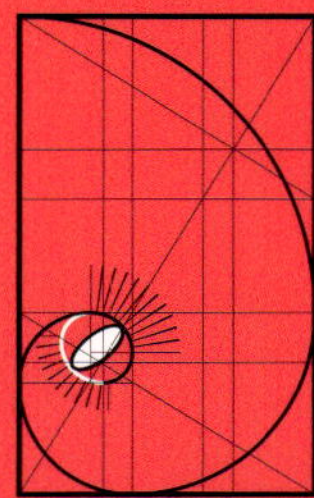
BEAUTY FALL 2023

LA+

BOTANIC SPRING 2024

LA+
INTERDISCIPLINARY JOURNAL
OF LANDSCAPE ARCHITECTURE

LA+

EXOTIQUE FALL 2024

LA+
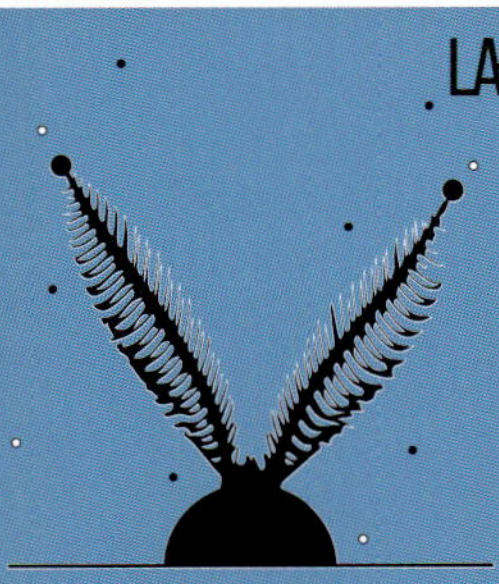
SENSE SPRING 2025

LA+

ENVIRONMENT FALL 2025

LA+

MEDIA SPRING 2026

LA+ brings you a rich collection of contemporary thinkers and designers in two issues each year. To subscribe follow the links at WWW.LAPLUSJOURNAL.COM